MW00526937

A Shadow in the Forest

A
Shadow in the
Forest

IDAHO'S BLACK BEAR

by

John J. Beecham

and

Jeff Rohlman

IDAHO DEPARTMENT OF FISH AND GAME

BOISE, IDAHO

AND THE UNIVERSITY OF IDAHO PRESS

MOSCOW, IDAHO

1994

Northwest Naturalist Books

Copyright © John J. Beecham and Jeff Rohlman, 1994
UNIVERSITY OF IDAHO PRESS,
Printed in the United States of America.

All rights reserved. No part of this publication may be repro-
duced, stored in a retrieval system, or transmitted in any form
or by any means, electronic, mechanical, photocopying, recording, or otherwise, except for purposes of scholarly review,
without the prior permission of the copyright owner.

Design by Caroline Hagen

98 97 96 95 94   5 4 3 2 1

Library of Congress Cataloging-in-Publication Data

Beecham, John.
    A shadow in the forest : Idaho's black bear / by John J.
Beecham and Jeff Rohlman.
        p.   cm. – (Northwest naturalist books)
    Includes bibliographical references and index.
    ISBN 0-89301-172-x
    1. Black bear – Idaho.   2. Mammal populations –
Idaho.   I. Rohlman, Jeff.   II. Idaho Dept. of Fish and
Game.   III. Title.   IV. Series.
QL737.C27B435   1994
599.74'446 – dc20
                                                94-1537
                                                CIP

THIS BOOK

IS DEDICATED TO MY MOTHER, ELINOR LUCILLE BEECHAM.
Her love of the outdoors and all of nature's creatures and her enthusiasm
for learning have been a constant source of inspiration.

John Beecham

CONTENTS

ILLUSTRATIONS

FIGURES

PHOTOGRAPHS

COLOR PLATES

TABLES

FOREWORD

BEARS HAVE LONG CAPTURED THE IMAGINATION OF humans. Likenesses of bears have appeared in earliest recorded history, attesting to a deep-seated interest, respect, and, perhaps, fear. Native peoples of North America learned to live with the great bears – the coastal brown bears, the inland grizzlies, and the awesome polar bears – and in some cases to revere them. Mutual respect seemed to order the sharing of a common homeland. While both humans and bears sought the same food, whether it be meat or plant, a workable system developed to use available resources usually without conflict.

Old-world settlers, however, had the firepower to kill bears, and they vigorously set about ridding the continent of ursid competitors. In many regions of the continent, all species of bears were literally wiped out. The adaptable black bear practically vanished from the eastern half of North America, and in the contiguous 48 states, the mighty grizzly was annihilated everywhere but in a few national parks and remote wilderness areas of the West. Even polar bears, despite the fact that they live in the most inhospitable regions on earth, were severely reduced in numbers across their entire circumpolar range.

The downward trend began to reverse in the mid-1900s for the black bear. The most adaptable of all the North American bears, it found some human practices to its liking. New plant growth following the cutting of mature forests, to cite one example, provided abundant food for bears in many regions. This, coupled with strict regulations on the killing of bears, resulted in increased populations.

International concern and alarm in the 1960s over diminishing polar bear numbers brought about action programs to conserve the great white bear. Cooperative international research and monitoring efforts, plus more restricitve hunting measures, have helped bring this bear back.

The grizzly, however, has not fared as well. In the 48 contiguous states, despite its official threatened and endangered status, grizzly numbers remain low, and the grizzly is confined to the Yellowstone and northwest Montana ecosystems. Human activities and development have resulted in habitat losses, with which the less adaptable grizzly simply cannot cope. On the positive side, population levels in the wilds of western Canada and Alaska are healthy.

The end to the downward spiral of bear numbers was brought about by public awareness and concern and the increased social value placed on all wildlife. However, achieving positive responses in populations and implementing practical management plans require information and knowledge. This knowledge, the essential foundation for any conservation strategy, comes from arduous research and investigation, and for long-lived species like bears, it must be long-term research if it is to be meaningful.

Over a long span of years, Dr. John Beecham and co-author Jeff Rohlman have performed that necessary research on Idaho's black bears. Their findings have played a major role in the successful management of regional black bear populations. Their work and the management program emanating from it is truly a wildlife success story.

We often hear that scientists talk only to scientists, that biologists only communicate with biologists. Sadly, this is too often the case, but this book is an exception. John Beecham and Jeff

Rohlman are to be applauded for making the effort to communicate their work to the public. It is a refreshing departure and one that I hope will be emulated by other agency biologists working not only with bears but other species as well.

Maurice Hornocker, Director
Hornocker Wildlife Research Institute

ACKNOWLEDGMENTS

THESE BLACK BEAR STUDIES WERE FUNDED BY IDAHO Federal Aid in Wildlife Restoration Project W-160-R, the Idaho Department of Fish and Game, the University of Idaho Cooperative Wildlife Research Unit, Montana State University, the University of Montana, and the Boone and Crockett Club of North America. In addition to these funding agencies and groups, a number of people contributed their time and energies to make these studies a success. Those that assisted us in the field were K. Ablin, J. Beecham, R. Beecham, C. Binninger, J. Brown, R. Davis, T. Ferguson, W. Greenberg, N. Johnson, J. Jonkel, M. Luque, K. Murphy, R. Myers, C. Nellis, A. Nicholson, S. Obenberger, A. Ogden, J. Pope, D. Rhodenbaugh, T. Rinkus, G. Servheen, M. Schlegel, R. Sherer, D. Wagner, and R. Wilmot. Graduate students that worked on various aspects of the project and contributed invaluably were S. Amstrup, D. Reynolds, J. Unsworth, and D. Young. C. Prentice and P. Wakkinen provided expert laboratory services. M. Hornocker, L. Irby, C. Jonkel, and L. Oldenburg provided administrative support, and J. Bryant provided editorial help. We offer a special thanks to the Dunning family for their support during our Priest Lake studies. B. T. Kelly assisted us with statistical analysis of growth patterns.

JOHN BEECHAM
JEFF ROHLMAN

I am not sure anyone can adequately acknowledge the support that comes from your family. However, a special thanks is due Jo Anne and Marnae and especially Denise for taking up the slack while I was in the field. Denise did a fine job of raising our children while balancing the demands of going to school fulltime, working, and taking care of everything else that came up during my absences. I also want to give a special thanks to my children, Jay, Blakely and Scott. They spent many nights sleeping in tents and fending for themselves while we were in the field. They set and maintained their own traplines around camp, collected bear scats, and helped record data on the bears we captured. Their companionship made the long absences tolerable and created fond memories for me.

JOHN BEECHAM

I owe special acknowledgment to my family, who supported me in my black bear research endeavors. I am grateful to my parents, Wendell and Carol Rohlman, for their financial and moral support for my research career. My wife, Erin, has my love and admiration for her support during my field excursions and her encouragement during the analysis phase of our black bear research in Idaho.

I also extend heartfelt appreciation and respect to John Beecham for accepting me into Idaho's black bear research program. He helped me advance in black bear research and in my wildlife management career. His insights as a mentor and as a friend will always be of great value to me.

JEFF ROHLMAN

ONE | INTRODUCTION

DID YOU KNOW THAT IDAHO'S BLACK BEARS ARE MORE active during the day than at night, that the bears' diet is less than 2 percent meat, and that the number of cubs born in the spring is related to the size of the fall berry crop? These are just a few of the facts we and other biologists with the Idaho Department of Fish and Game have gathered as part of our black bear research program. Since 1972, we have been collecting the biological data needed to develop a comprehensive management program for Idaho's black bears. This book summarizes the results of our black bear research. Although we are Department of Fish and Game employees, the opinions expressed in this book are our own and not necessarily those of the department.

Gathering data on Idaho's bears hasn't been easy. The black bear (*Ursus americanus* Pallas) is a shy, adaptable species whose distribution in Idaho (Fig. 2–3) coincides closely with that of the state's coniferous forests. The black bear's secretive habits and preference for forested habitats make it a difficult animal to observe, for both scientists and the public. Prior to the early 1970s, when we began using *radio-telemetry equipment* on bears, most of the information we had on their biology was from occasional descriptions of bear sightings. By 1990, we had conducted hundreds of black bear research studies throughout the bear's range, significantly improving our ability to understand this species and its habitat requirements.

Although the black bear is widely distributed in Idaho, it has never occurred in high densities (number of animals per square mile). In spite of this, the black bear is a popular big game animal pursued by hunters in the spring and fall, and it is increasingly becoming an important species to the outfitting industry. The most important documents attesting to the recognized value and status of black bears in Idaho are the state's Black Bear Species Management Plans, which establish the Idaho Department of Fish and Game's philosophy and management direction for bears. This book doesn't include those management plans, because they change frequently. For a current version of the black bear management plan, contact the department. The plans, which are based on input from a variety of groups interested in black bear management, reflect the positive attitude the people of Idaho have toward bear management and the strides we have made in managing this species.

To date, we have studied six geographically separate bear populations (Fig. 1–1). Initially we designed these studies to learn about the status of each black bear population, although we also collected data on bear food habits, physical condition, denning requirements, and behavior. During the past six years, our emphasis shifted toward learning more about black bear habitat use patterns and developing a population monitoring system.

Our field research on Idaho black bears began in 1973 when we initiated a five-year population ecology study near Council in west-central Idaho. Our objectives were to learn the population size, sex and age structure, movement and activity patterns, reproductive biology, denning activities and den characteristics, and food habits of a hunted bear population. In 1975, the hunting season in this study area was closed; during 1982 we resampled this population to see if any population characteristics had changed. We also placed ten radio collars on adult female bears to monitor their habitat use and reproductive performance.

In 1984, we began another research effort at Council to develop

Figure 1-1. Location of black bear ecological study areas in Idaho.

The development of radio-tracking equipment made it much easier to study Idaho black bears. Photo by John J. Beecham

a population monitoring technique, which we successfully concluded in 1990.

From 1975 to 1979, we conducted a black bear study in north-central Idaho, near Lowell, in conjunction with a calf elk mortality study being done by department biologist Mike Schlegel, to determine the size, sex composition, and age composition of an essentially unhunted bear population. We also wanted to measure the response of this population to the major bear relocation program done as part of the calf elk mortality study.

During 1978, we studied a third population in the Coeur d'Alene River drainage north of Wallace, and collected data on sex and age structure, population size, and food habits. The bear hunting season was liberalized in 1980, so we sampled this study area again in 1983 to see how the bear population responded to increased hunting pressure.

We conducted a fourth study between 1979 and 1981 on the east side of Priest Lake. Our objectives were essentially the same as those for the Council study, except that we placed more emphasis on black bear habitat use patterns in our telemetry studies. From 1984 to 1990, we also used Priest Lake as a test area for our population monitoring research.

We conducted single-year studies near Elk River and along the St. Joe River in 1981 and 1982, respectively. To learn the population status of bears in those areas, we collected data on sex and age distribution, population size, and food habits.

After completing these six studies, we saw that we still needed a method to estimate population size or trends without conducting expensive capture/release studies. So in 1984, we initiated a project designed to measure black bear population trends in several of our study areas (Council, Priest Lake, Coeur d'Alene, and St. Joe River). The bait station technique we used was originally developed in Tennessee, but we modified the procedure to fit Idaho's diverse habitats. We continued the research through the summer of 1990.

In this book, we focused on the Council, Priest Lake, and Lowell studies because they have larger databases. We also presented data collected for the Coeur d'Alene, Elk River, and St. Joe River studies, as well as data from black bear ecology research in other areas of North America. Much of this has already been published in technical journals, but it is repeated here to make our information available to readers who might have an avid interest in and desire to learn more about Idaho's black bears.

Numbers in parentheses in the text, e.g., (1), are keyed to a list of references at the end of each chapter. Notes appear in the text as superscript numbers, e.g.,[1], and are keyed to an endnote section at the end of a chapter.

TWO | EVOLUTIONARY HISTORY, CLASSIFICATION, AND DISTRIBUTION

THE BEAR FAMILY ORIGINATED MORE THAN 20 MILLION years ago in the Palearctic region (Europe, northern Asia, northern Arabia, and Africa north of the Sahara). From these beginnings the black bears evolved. During the Pleistocene era, about 100,000 years ago, black bear ancestors moved across the Bering land bridge to North America. In recent years the black bear's range in North America has shrunk, but the bear can still be found in most states and Canadian provinces (Fig. 2–1).

EVOLUTIONARY HISTORY AND CLASSIFICATION

Kingdom: Animal	Family: Ursidae
Phylum: Chordata	Genus: *Ursus*
Order: Carnivora	Species: *americanus*
Suborder: Fissipeda	Common Name: American black bear

Bears (ursids) are most closely related to the canid (dog) family and geologically are the youngest of all the carnivores. Fourteen genera of bears have been identified (five extant; nine extinct). Those bears that survive in today's world include the *Ursus* genus, which has four species, and the *Ailuropoda* (panda bear), *Helarctos* (sun bear), *Melursus* (sloth bear), and *Tremarctos* (spectacled bear) genera, which have only one species each. Ursids evolved from miacids, a

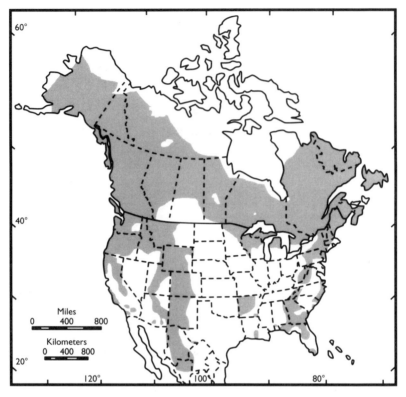

Figure 2–1. Distribution of the black bear (*Ursus americanus*) in North America – modified from Pelton (1982). Used with permission of Michael R. Pelton and The Johns Hopkins University Press.

family of small, carnivorous tree-climbing mammals (1), some 20 to 25 million years ago in the Palearctic region. The earliest members of the ursid family were forest dwellers who eventually came to depend on a diet of largely vegetable matter. When the opportunity presented itself, they also fed on animals.

The first member of the bear family identified from the fossil record was *Ursavus* (Fig. 2–2). From *Ursavus* came two lines of bears: the Tremarctinae, which has only one surviving member, the spectacled bear, and a second line that evolved into the Ursinae line, ancestor to the black bear. The earliest member of the Ursinae line was *U. minimus*, a forest dweller who resembled the American black

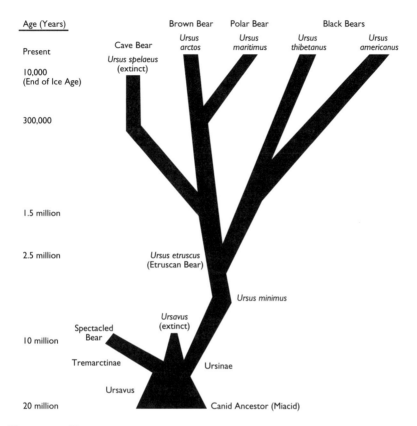

Figure 2–2. *Ursus* genus.

bear, although *U. minimus* was smaller. *U. minimus* eventually evolved into the Etruscan bear (1), which then split into three lines. One of these lines, the cave bears of Europe, became extinct near the end of the last ice age some 10,000 years ago. The other two lines led to the black bears and brown bears. The polar bear evolved from the brown bear line about 300,000 years ago.

DISTRIBUTION

All descendants of the ursid line evolved in the Palearctic region. The three species found in North America – the black, brown, and polar bears – are all immigrants that moved onto the continent during the Pleistocene era. The brown bear group, which includes a subspecies known as the grizzly bear, has a worldwide distribution in the Northern Hemisphere. The brown bear's ancestors arrived in North America sometime within the past 100,000 years. The black bear is found only on the North American continent. Historically, its range included all the forested areas of the continent from Alaska to the northern states of Mexico and from California east to Florida and the Canadian provinces of Newfoundland and Nova Scotia (2). The black bear was apparently absent only from the Central Plains states and provinces, and the barren grounds of northern Canada and Alaska. The oldest fossil remains of black bears were found in Pennsylvania and are estimated to be 500,000 years old.

Today, the black bear's range has shrunk, but still includes all or parts of 38 states, 11 Canadian provinces, and 7 Mexican states. Jonkel and Miller (3) reported that the black bear has been observed with increasing frequency on the barren grounds of Canada. They speculated that this apparent range extension was related to diminishing numbers of grizzly bears in the area. The reduction of the black bear throughout the remainder of its range appears to be related to habitat loss from the encroachment of man and his activities.

In Idaho, most bears inhabit the coniferous forests of the state (Fig. 2–3). North of the Snake River Plain, black bears are found throughout the forested mountains and foothills, and few black bears occur south of the Snake River, except in southeastern Idaho. About 75 percent of the black bear habitat in Idaho is administered by the U.S. Forest Service, 20 percent is controlled by private interests, and the rest is administered by other agencies, such as the Bureau of Land Management (BLM), the Idaho Department of Lands (IDL), and the Idaho Department of Fish and Game.

Although no data are available on the number of black bears in Idaho, we applied our research data on black bear densities in specific study areas to get an estimate of the maximum number of bears the available habitat could support. Based on our 1990 calculations, Idaho's 30,000 square miles of habitat could support a maximum of 20,000 to 25,000 black bears. However, research we conducted along

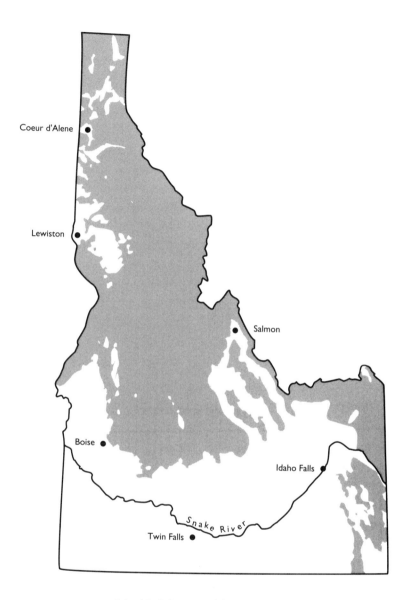

Figure 2–3. Range of the black bear in Idaho.

the Coeur d'Alene and St. Joe rivers and in the Elk River area showed lower bear densities than we expected based on habitat quality. This leads us to conclude that the actual number of black bears in the state in 1990 was probably below the maximum the habitat can support.

LITERATURE CITED

1. HERRERO, S. 1972.
   Aspects of evolution and adaptation in American black bears and brown and grizzly bears of North America. Int. Conf. Bear Res. and Manage. 2:221–31.
2. COWAN, I. MCT. 1972.
   The status and conservation of bears of the world – 1970. Int. Conf. Bear Res. and Manage. 2:343–67.
3. JONKEL, C. J., AND F. L. MILLER. 1970.
   Recent records of black bears on the barren grounds of Canada. J. Mammal. 51(4):826–28.

THREE | STUDY AREAS

BECAUSE BLACK BEARS ARE SECRETIVE AND DON'T occur in high numbers, selecting study areas was one of the most stressful parts of our research program. The logistics of establishing a base camp and building 30 to 40 trap sites were significant and not taken lightly because of the limited amount of time and money we had to trap bears each summer. Some of the factors we considered in choosing study areas were habitat quality, accessibility, elevation, and whether the area contained a mixture of undisturbed and *disturbed vegetation types*. We also looked at other activities in the area that might affect our trapping efforts, the level of hunting pressure, and whether our wildlife managers were concerned about the status of a bear population in a given area.

Selecting our first study area near Council was especially difficult. I knew the area was popular with bear hunters and assumed it contained good bear habitat. The area had been heavily logged and had many roads, which meant we would have plenty of access for trapping bears, but also that there were few escape areas for bears during hunting season. Would hunting mortality rates be too high and interfere with our studies? Council was also close to the heavily populated Treasure Valley and many people who enjoyed fishing, picnicking, woodcutting, and other activities. Would these activities influence our studies?

As it turned out, bear hunting pressure in our Council study area was excessive, and the Department of Fish and Game eventually closed the bear season to protect the radio-collared female bears we were studying. The large number of summer visitors had a minimal impact on our trapping success. In fact, we were able to get several visitors involved in our trapping program. I'm sure some of them took home a new respect for bears because of their contact with black bears and our research team.

As time went on, selecting a study area became easier. In June 1979, as I was finishing up my last field season at Lowell, I sent Jeff Rohlman and Kris Ablin to Priest Lake to select a study area and begin building trap sites. I planned to join them in early July. I heard little from Jeff or Kris that month and knew they must be feeling some of the same pressures I felt while choosing a study area at Council in 1973. I also knew they were experienced bear trappers and good biologists, so I was confident they could do the job. When I arrived at Priest Lake just after the 4th of July, I was in for the surprise of my life. Not only had Jeff and Kris picked one of the most beautiful campsites I could imagine, they had also built more than 20 trap sites and caught 41 bears – all in the first month. I knew then that our experience with bears was paying off and that selecting future study areas would be much less stressful.

JOHN BEECHAM

COUNCIL

This study area, located in west-central Idaho about 10 miles southeast of Council, encompasses nearly 50 square miles. The predominant geographic features are Council Mountain, West Mountain Ridge, and the Middle Fork of the Weiser River. Council Mountain is characterized by strongly glaciated lands, while West Mountain Ridge has gently rolling, *periglaciated* slopes with few steep inclines. The Columbia River basalt formation, with its fine-textured soils, and the Idaho batholith granite formation, with its coarse-textured soils, are the major geologic types. Elevations range from 3,200 to over 8,100 feet on Council Mountain, and slopes generally exceed 30 percent.

Logging is a major land use activity in the Council, Coeur d'Alene, Priest Lake, Elk River, and St. Joe study areas. Photo by Jeff Rohlman

Although the climate of the area is continental, it is strongly influenced by air from the Pacific Ocean. The area has moderately long, cold, wet winters and hot, dry summers. Annual precipitation ranges from 26 inches at lower elevations to 45 inches at higher elevations and averages 32 inches. The mean annual temperature at Council is 48°F.

Big sage, grasses, and *forbs* dominate the vegetative communities at lower elevations and on drier sites. Timber stands range from dense to open with many naturally occurring, thin-soiled areas or "scablands" that have little or no vegetative cover. At lower elevations (less than 5,575 feet), timber stands are confined to *riparian* areas and north-facing slopes, and are dominated by ponderosa pine and Douglas fir. Hawthorn, chokecherry, and elderberry are common in the riparian zones. At middle and higher elevations, grand fir, subalpine fir, and Engelmann spruce are the dominant tree species. Important understory shrubs are huckleberry, buffaloberry, bitter cherry, dogwood, and mountain ash. Whitebark pine occurs on some ridgetops and lodgepole pine occurs on some disturbed areas.

The major land uses affecting the Council area are commercial timber cutting and livestock grazing. Logging on public lands began in the 1950s, and from 1960 to 1970 loggers removed over 225 million board feet of timber from the Council Ranger District. About 1,000 cows and calves graze the study area (the grazing season on U.S. Forest Service lands runs from July until mid-October).

In the summer, recreationists fish for trout in the Middle Fork of the Weiser River and its tributaries, camp, picnic, and pick berries. Most fall recreational activities center around the mule deer hunting season, and a short elk hunting season in early October. In 1975, the black bear hunting season in the Middle Fork and Little Weiser drainages was closed to protect radio-collared female bears in the study area. The bear season was opened during the general deer season from 1980 to 1982, but was closed again when we began studying black bear population trends using bait stations.

LOWELL

This study area encompasses 100 square miles about 100 miles east of Lewiston. The main geographic features are Coolwater Ridge, which bisects the study area from west to east, the Lochsa River, which borders the area on the north, and the Selway River, which is located on the south side of the study area. In most years, motorists can access the area only from July through October on a single road along Coolwater Ridge. Elevations range from 1,510 to 7,000 feet.

Pacific Ocean air currents strongly affect the climate of the area, which is continental. This causes moderately long, cold, wet winters and hot, dry summers. Annual precipitation averages 32 inches. Timber stands at lower elevations are relatively dense and contain ponderosa pine, Douglas fir, and western larch. Although western red cedar dominates most timber stands at midelevations, grand fir, Douglas fir, western larch, and white pine are also present. The Lowell area has a long history of wildfires dating back to the 1750s and, as a result, contains vast areas of *seral* brushfields in the red cedar zone primarily composed of willow, maple, service-berry, bitter cherry, mountain ash, ceanothus, twinberry, syringa, and ocean spray. At higher elevations subalpine fir is the major tree species. Engelmann spruce and lodgepole pine, two seral species, also grow in this zone. Wet and dry meadows are common in the spruce-fir zone, and bears use them extensively for feeding during the spring. Important understory species occurring in the area include grasses, horsetail, clover, licoriceroot, strawberry, American false hellebore, huckleberry, and snowberry.

Moderate recreational use, primarily hunting for elk and bear, occurs on the area, and several hundred cattle graze here during the summer.

COEUR D'ALENE

This study area, located in the Coeur d'Alene River drainage, is about 50 square miles in size. Elevations range from 2,600 to 6,200 feet.

Airflows from the Pacific Ocean influence the climate, which is characterized by moderately long, cold, wet winters and dry summers. Precipitation averages 47 inches per year, most of it falling as snow. The mean annual temperature is 48°F.

The three major vegetative zones on the area are (1) the pine-fir zone between 2,600 and 2,800 feet, (2) the cedar-hemlock zone between 2,800 and 4,900 feet, and (3) the spruce-fir zone above 4,900 feet (1). In these zones, major tree species include subalpine fir, western red cedar, Engelmann spruce, lodgepole pine, white pine, and Douglas fir. Predominant understory species are fool's huckleberry on the wetter sites, huckleberry at middle and higher elevations, and beargrass on exposed (drier) sites.

The Coeur d'Alene National Forest (now part of the Panhandle National Forests) has a long history of mining and logging activities dating back to the 1800s. During the 1950s, logging became a significant influence on the forest when loggers *clear-cut* large areas of timber (1). Roads constructed during these logging operations now provide easy access throughout the study area. The Yellowdog and Flat Creek tributaries were heavily logged, with clear-cuts covering about 23 percent of the area in 1978. The Brett and Miners Creek tributaries weren't as heavily logged, with clear-cuts covering only 5 percent of the area. The Coeur d'Alene study area experiences moderate recreational use consisting of fishing, picnicking, and berry picking during the summer months and elk hunting in the fall.

PRIEST LAKE

This 115-square-mile study area is located within Bonner and Boundary counties along the east side of Priest Lake. Elevations range from 2,300 to 7,600 feet. The topography is steep and rugged, with exposed bedrock common above 5,500 feet.

Pacific maritime air currents affect the climate, resulting in long, snowy winters and short, damp summers. Annual precipitation averages 32 inches and the mean annual temperature is 41°F.

The western hemlock *potential climax series* (2) dominates forests at lower elevations (less than 3,900 feet) and middle elevations (3,900 to 5,200 feet). Pockets of the Douglas fir series occur on dry sites at lower elevations, while the western red cedar series grows on wetter sites at lower to middle elevations. At higher elevations (greater than 5,200 feet), the subalpine fir series is dominant, with the subalpine fir/whitebark pine series occurring above 5,900 feet. The interspersion of logged areas, burns and *sidehill parks* throughout these forests creates a diverse mosaic of plant communities.

Most lands in the Priest Lake study area are administered by the Idaho Department of Lands. Timber production has been the major land use practice affecting the study area. During the mid-1940s, commercial logging began in the southern Selkirk Mountains. Timber harvest proceeded northward until, by the 1960s, loggers had removed most timber in the northern areas. Since the 1940s, parts of the Priest Lake study area have been commercially logged several times. Initially, loggers *select-cut* white pine for saw logs. In subsequent harvests, they cut a variety of tree species for poles or pulpwood.

Other land use practices at Priest Lake include recreational activities such as backpacking, berry picking, camping, hunting, fishing, firewood cutting, and snowmobiling.

One of our bear study areas was located on the east side of Priest Lake. Photo by Jeff Rohlman

Historically, wildfire was a significant factor affecting the Selkirk landscape, although effective fire suppression during the past 50 years has reduced this influence (3). However, in 1967, two large wildfires occurred in the Priest Lake area: the 25,250-acre Sundance burn bordering the study area to the south, and the 10,850-acre Trapper Peak burn 11 miles north of the study area.

ELK RIVER

The Elk River study area is located about 41 miles east of Moscow and is 25 square miles in size. The predominant geographic features include Shattuck Butte, Elk Butte, and Windy Point. The major drainages are Johnson Creek, Shite Creek, Elk Creek, and Morris Creek. Elevations range from 2,900 to 5,870 feet. The climate in this area is similar to that described for the Lowell study area.

The Elk River study area contains a mosaic of natural openings, mature timber stands, and logged areas. Mature timber stands vary from the Douglas fir/ninebark *habitat type* at lower elevations to

the subalpine fir/mountain lover habitat type at the highest elevations.

Logging began in the Elk River area as early as 1910, but extensive logging didn't occur until the 1960s. From 1967 to 1976, loggers
removed 15,175 board feet per acre per year, while the average
amount of timber produced from 1964 to 1974 was 505 board feet per
acre per year. Roads constructed during these logging efforts provide easy access into the area.

Other major land uses include grazing and recreation. Grazing occurs in many of the natural meadows and in some of the more
recent clear-cuts and burns. Recreational uses include fishing,
hunting, hiking, and snowmobiling.

ST. JOE

This 58-square-mile study area is located about 47 miles southeast
of Coeur d'Alene, centered in the Marble Creek drainage on the
south side of the St. Joe River. Although Marble Creek has been
heavily roaded, it contains some reasonably large blocks of unroaded
terrain, so we considered it representative of the St. Joe River drainage. Elevations range from 2,300 feet along the St. Joe River to over
5,900 feet.

Although somewhat drier than the Coeur d'Alene study area,
this area has a climate characterized by cold, wet winters and dry
summers.

Area vegetation is similar to that found on the Lowell and
Coeur d'Alene study areas, including the large seral brushfields
created by wildfires during the early part of this century.

Logging is the primary land use practice affecting this heavily
roaded area. Livestock and recreational use is light and seasonal.

LITERATURE CITED

1. IRWIN, L. 1978.
Relationships between intensive timber cultures, big game habitat,
and elk habitat use patterns in northern Idaho. Ph.D. dissertation,
the University of Idaho, Moscow. 282pp.

2. DAUBENMIRE, R., AND J. DAUBENMIRE. 1968.
Forest vegetation of eastern Washington and northern Idaho. Wash. Agric. Exp. Stn. Tech. Bull. 60. 104pp.

3. ZAGER, P. E. 1981.
Northern Selkirk Mountains grizzly bear habitat survey, 1981. U.S. For. Serv., Idaho Panhandle Natl. For. Contract. 75pp.

FOUR | METHODS

BEFORE 1973, VERY FEW IN-DEPTH STUDIES OF BLACK bear ecology had been attempted. Most of our knowledge about bears came from observations of bears at zoos, in parks, or in laboratory settings. As a result, we had to develop many of our own techniques for studying free-ranging black bears in Idaho.

For example, there was some information in the scientific literature on when and where black bears denned, but no details on how to collect this information. We knew our radio-collared bears would eventually lead us to their winter dens, but how could we approach the bears without causing them to abandon their dens? How would our tranquilizing drugs work on hibernating bears? Would the bears be aggressive? Would we need to carry a gun for protection? Would we even be able to find the den location under several feet of snow?

The first den we approached in the winter of 1973–1974 was occupied by No. U-41, an adult female with two yearling cubs. Locating the den was easy. Getting into the den and successfully tranquilizing the three bears, however, was a challenge. The den had a four-foot-long tunnel leading into the den chamber. The tunnel was about 20 inches in diameter, so I didn't have enough room to hold a flashlight and maneuver the *jab stick* to inject the bears with tranquilizing drugs. Fortunately, No. U-41 and her cubs weren't particularly aggressive, and I was able to hold the flashlight and guide the jab stick with my fingertips while one of my assistants pushed the stick into the den from behind me. The tranquilizing drugs worked fine, but it did take longer than usual for the bears to go out because the bears' body metabolism was slower than during the summer.

After we removed the bears from the den and replaced the radio collar on No. U-41, we hit a major, unexpected snag. It was extremely hard to put No. U-41 back in her den. In fact, it was like trying to push 125 pounds of Jell-O (without a bowl) though a 20-inch hole. It couldn't be done. We finally solved our dilemma by having one researcher enter the den and pull on the bears from the inside while the rest of us pushed from the outside. Once the bears were settled comfortably, the researcher crawled out, with a little help from us.

During the course of our studies, we entered more than 100 dens and tranquilized their occupants. Each den presented new challenges, and our methods evolved continually to meet those challenges.

JOHN BEECHAM

As we conducted our bear studies, we used standard procedures developed by other researchers when possible, such as radio-tracking and vegetation-measuring techniques. When no standard method existed, we had to improvise. We used the following methods for trapping and handling bears, radio-collaring bears (telemetry), sampling bear habitats, examining denned bears, and collecting bear harvest data.

TRAPPING AND HANDLING

We captured black bears using Aldrich spring-activated foot snares set in or adjacent to *cubby sets* (see Fig. 4–1) or on trails leading to these sets. Snares were attached to "green" drag logs or living trees next to the cubbies and trails. During 1973 and 1974, we used a variety of baits, including pig and cattle heads, carp, road-killed deer,

Figure 4–1. Typical cubby set used to trap black bears. Photo by John J. Beecham

and cattle offal, to attract bears to the cubbies. After 1974, we were able to get spawned-out salmon and steelhead carcasses from department hatcheries, so we used them almost exclusively.

We immobilized snared bears with intramuscular injections of Sernylan[1] and Sparine[2] during the early years of the study (1973 to 1978). After 1978, Sernylan became very difficult to obtain, so we began using a combination of Ketaset[3] and Rompun.[4] We administered drugs to most snared black bears with a syringe mounted on the end of a six-foot jab stick (Plate 1). In 1975, on the Lowell study area, we darted 12 adult black bears from a Hiller 12-E helicopter, using a Cap-chur gun and a standard 300-milligram dose of Sernylan.

After the bears were immobilized, we tagged them with numbered aluminum or plastic ear tags and tattooed the same number in their right ear and upper lip. We took physical measurements, blood samples, and physiological data from most bears and extracted a tooth (I_1 or P_1) from bears older than two years. The teeth were aged using the cementum-annuli aging technique (1) (Fig. 4-2).

We examined the mammary glands and vulvas of female bears to determine their reproductive status (2). To provide a rough indicator of male reproductive status, we measured the bears' testes (through the scrotal sac) to the nearest millimeter for length and width.[5]

Figure 4–2. Cross section of the root of an incisor tooth showing annuli. Photo by John J. Beecham

With the exception of 75 bears captured at Lowell in 1976, we released all bears, after handling, at their capture location. The Lowell bears caught in 1976 were transported by helicopter from their capture site to a small meadow along the Selway River using a cargo net attached beneath the helicopter. We then placed 72 of these bears in holding cages and transported them to one of 12 Idaho release locations chosen in consultation with U.S. Forest Service district rangers. To minimize the impact of relocating 72 bears, we released them in groups of 3 to 11 animals. Family groups were held in a single cage and released intact.

TELEMETRY

Forty-six bears were radio-collared on the Council study area and eight bears on the Priest Lake study area. We monitored the bears from the ground and from an airplane to learn their movements, activity periods, denning locations, and habitat use patterns. We plotted their locations on U.S. Geological Survey topographic maps.[6]

To determine their daily activity patterns, we periodically monitored selected bears at hourly intervals for 24-hour periods. Their activities were recorded as (1) bedded, (2) feeding, (3) traveling, (4) denned, or (5) unknown. Radio collars purchased after 1978 were equipped with motion-sensitive devices that changed the transmitter pulse to a slow mode (50 to 60 beats per minute [bpm]) if the collars stayed motionless for two minutes. The pulse remained fast (75 to 85 bpm) if the animal was active. We determined the bears' activities by monitoring bear pulse rates, observing bears, or looking for bear sign (scats, tracks, etc.) in the areas the bears had been using. If there was a significant change in the direction of a signal during the location process, we considered a bear to be traveling.

We learned habitat use patterns on the Council and Priest Lake study areas by monitoring bears from the ground and classifying their locations as (1) visual – the bear was seen; (2) close – the bear was within 100 yards, based on radio signal strength or hearing the bear without *triangulation*; (3) close triangulation – the bear was within 300 yards, based on radio signal strength and triangulation;

and (4) triangulation – the bear was farther than 300 yards away, based on triangulation. We used close triangulation and triangulation infrequently[7] (3, 4).

HABITAT SAMPLING AT BEAR LOCATIONS

To classify black bear habitats, we modified a habitat component hierarchy developed by the University of Montana Border Grizzly Project (5, 6) and Zager (7). We used both *cover types* and habitat types (8) to describe bear habitats. Our habitat classification system also provided a systematic and repeatable means of classifying non-forested areas, such as burns, meadows, *scree*, and seral plant communities used by black bears.

We roughly delineated cover types from aerial photographs and then verified them on the ground. Using 0.1-acre circular plots, we randomly sampled the vegetation in cover types. The major plant species found in the overstory and understory were recorded and given a cover class value (9) based on the amount of horizontal area they covered.[8] To determine vertical cover, we visually estimated the percent cover in each of four strata[9] in sample plots.[10] We recorded the topography of cover types as ridgetop, upper slope, midslope, lower slope, bench or flat, or stream bottom. The horizontal configuration of cover types was classified as convex (dry), straight (level), concave (wet), or undulating.

We employed a nonmapping, random dot grid technique (10) to determine the availability of all cover types within the bears' composite home range.

Cover type selection patterns were developed entirely from the radio locations of collared bears. We didn't use incidental bear sign in the analysis because of the inherent biases they bring to the data set. A cover type was considered "selected for" if bears used it significantly more than expected and "selected against" if they used it significantly less than expected, based on the availability data.[11]

Each field season, we recorded the developmental stages (budding, flowering, etc.) of plants identified as important bear foods on the study areas (11, 12, 13, 2). We studied permanent plots located at varying elevations and aspects, in areas used by bears and picked at random, using a modification of the method described by West and Wein (14). We identified key bear food plants from scat analyses conducted on all the study areas. Statistical analysis[12] was used to measure the degree of correlation between the mean elevations of bear locations and the developmental stages of key bear food plants.

DENNING

We found bear dens by monitoring the activity patterns and locations of radio-collared bears in October and November. When a bear showed up repeatedly at a particular site and its activity level dropped significantly, we assumed it was denned. We then quietly approached the site and marked it by attaching surveyor's tape to nearby trees.

We visited most dens in December or March to remove or replace old transmitter collars and to record the physical characteristics of each den and den site. Snowmobiles and Cushman Tracksters were used to approach each den as closely as possible. We then backpacked our equipment to the den site on snowshoes.

After shoveling the accumulated snow from the den entrance, we immobilized the occupants and removed them from the den for weighing and measuring. We also recorded the average height and width of den entrances; the height, width, and length of den tunnels and chambers; the ground slope in degrees; the percent canopy of coniferous trees over the den; den aspect; and den elevation. After we finished, we returned the bears to their dens and covered the entrances with snow. Several statistical methods were used to compare denning data.[13]

HARVEST DATA

We relied on two methods to collect black bear harvest information: (1) the mandatory check and report program and (2) the annual telephone harvest survey. The mandatory check and report program, begun in 1983, required a hunter to bring the skull of his/her harvested bear to an official checkpoint within ten days of killing the bear and to fill out a report form. In most cases, we extracted a tooth from the skull for aging purposes. Pertinent data, including the bear's sex, kill date, location of kill, and method of take, were recorded.

The annual telephone harvest survey provided a second estimate of the black bear harvest. The telephone survey crew contacted about 3 percent of the bear tag holders and collected information from both successful and unsuccessful hunters. From this information, we could estimate hunter success and the number of recreation days men and women spent hunting black bears each year.

LITERATURE CITED

1. STONEBERG, R. P., AND C. J. JONKEL. 1966.
   Age determination of black bears by cementum layers. J. Wildl. Manage. 30(2):411–14.
2. REYNOLDS, D. G., AND J. J. BEECHAM. 1980.
   Home range activities and reproduction of black bears in west central Idaho. Int. Conf. Bear Res. and Manage. 4:181–90.
3. YOUNG, D. D., AND J. J. BEECHAM. 1986.
   Black bear habitat use at Priest Lake, Idaho. Int. Conf. Bear Res. and Manage. 6:73–80.
4. UNSWORTH, J. W., J. J. BEECHAM, AND L. R. IRBY. 1989.
   Black bear habitat use in west central Idaho. J. Wildl. Manage. 53:668–73.
5. ZAGER, P. E., C. JONKEL, AND R. MACE. 1980.
   Grizzly bear habitat terminology. Border Grizzly Proj., University of Montana, Missoula, Spec. Rep. No. 41. 15pp.
6. SERVHEEN, C. 1981.
   Grizzly bear ecology and management in the Mission Mountains, Montana. Ph.D. dissertation, University of Montana, Missoula. 139pp.
7. ZAGER, P. E. 1981.
   Northern Selkirk Mountains grizzly bear habitat survey, 1981. U.S. For. Serv., Idaho Panhandle Natl. For. Contract. 75pp.
8. DAUBENMIRE, R., AND J. DAUBENMIRE. 1968.
   Forest vegetation of eastern Washington and northern Idaho. Wash. Agric. Exp. Stn. Tech. Bull. 60. 104pp.

9. PFISTER, R., B. KOVALCHIK, S. ARNO, AND R. PRESBY. 1977. Forest habitat types of Montana. U.S. For. Serv. Gen. Tech. Rep. INT-34. 174pp.

10. MARCUM, C. L., AND D. O. LOFTSGAARDEN. 1980. A nonmapping technique for studying habitat preferences. J. Wildl. Manage. 44:963–68.

11. AMSTRUP, S. C., AND J. J. BEECHAM. 1976. Activity patterns of radio-collared black bears in Idaho. J. Wildl. Manage. 40:340–48.

12. BEECHAM, J. 1976. Black bear ecology. Job Prog. Rep. Idaho Dept. Fish and Game, Boise. 34pp.

13. BEECHAM, J. 1977. Black bear ecology. Job Prog. Rep. Idaho Dept. Fish and Game, Boise, 43pp.

14. WEST, N. E., AND R. W. WEIN. 1971. A plant phenological index. BioScience 21(3):116–17.

ENDNOTES

[1] Phencyclidine hydrochloride; 0.6 milligram/pound of body weight.
[2] Promazine hydrochloride; 0.3 milligram/pound of body weight.
[3] Ketamine hydrochloride; 2.0 milligram/pound of body weight.
[4] Xylazine hydrochloride; 1.0 milligram/pound of body weight.
[5] Testis size was expressed as an equivalent diameter (ED) where ED = length + width ÷ 2. We used maximum mean testis size for males three years of age or older to indicate the peak of the breeding season. To predict the age of first reproduction for Council, Lowell, and Priest Lake female black bears, we employed segmented regression analysis to attempt to identify the point of maximum curvature on their growth curves and to compute separate linear regression equations for the pre- and postinflection phases of the curve. We used covariance analysis (ANCOVA) to test for differences in the slopes of these regression lines between each population.
[6] Map scale was 1:62,500.

[7] Close triangulation and triangulation were included in the habitat use analysis only if all compass bearings intersected at a single location within a large uniform cover type.

[8] Cover class values were $1 = 0$ to 1.9 percent, $2 = 2$ to 5 percent, $3 = 6$ to 25 percent, $4 = 26$ to 50 percent, $5 = 51$ to 75 percent, $6 = 76$ to 95 percent, and $7 = 96$ to 100 percent.

[9] The strata used were zero to 2 feet, 2 to 7 feet, 7 to 25 feet, and greater than 25 feet.

[10] Differences in vertical cover between cover types were detected using a grouped t-test ($P \leq 0.10$).

[11] The chi-square goodness of fit test was used to detect significant differences ($P \leq 0.10$) between the availability and use of cover types. To see if bears preferred or avoided individual cover types, we applied a modified Z statistic (10).

[12] Simple linear regression.

[13] Because of small sample sizes, we used the Wilcoxin Rank Sum test to compare the dates that males and females entered their respective dens. The Student's *t* test was used for other statistical comparisons of denning data.

FIVE | MORPHOLOGICAL AND PHYSIOLOGICAL CHARACTERISTICS

SINCE 1972, WHEN I FIRST BECAME INVOLVED IN black bear research, I have discussed bear biology with thousands of people and have found that nearly everyone had a bear story to tell. In many cases, the stories were about firsthand experiences; in others, they involved encounters between a bear and a close friend. In almost all cases, though, the bear was huge! Bear weight estimates of 400 pounds or more weren't uncommon and seemed to indicate that our bear population was extremely well nourished.

In this chapter on bear morphology and physiology, one of the subjects we'll discuss is the size of the black bears we captured here in Idaho. But how do those bears compare to the one you saw last year or to a bear you might shoot next year? Judging the size of a bear is difficult. During our trapping season, we often had contests to see who could come closest to guessing a bear's actual weight. On many occasions, we missed the bear's weight by 30 to 50 pounds – and we were standing just 15 feet away, had a good, long look at the bear, and had weighed hundreds of bears before. It's no wonder that the average person, who sees very few bears in a lifetime, can misjudge a bear's size under the best of circumstances.

There's not much we can do to help you accurately judge

the size of the next bear you see in the wild. We can tell you, however, that the average adult female (adults are four years or older) will weigh 120 to 140 pounds during the summer and the average adult male will weigh twice that.

We can provide a little more aid to hunters, who may get to examine a bear up close. We weighed and took several body measurements of the bears we captured each year, then compared how useful each body measurement was in predicting a bear's weight. None of the measurements were perfect, but a bear's neck circumference was the best predictor of its body weight (Table 5–1). Measurement errors, differences in pelt thickness or body configuration, and other factors caused variations in body weight estimates for a given neck size. However, by measuring the neck circumference of the next bear you shoot, you can come up with a reasonable estimate of how much it weighed. (Caution: make sure the bear is *Dead* before you use this technique.) For example, using Table 5–1 as a guide, if you shoot a male black bear with a 27-inch neck circumference, you can be 90 percent sure that the bear weighed between 217 and 285 pounds (251 + 34 pounds). Although this isn't a perfect solution to the question of bear size, a tape measure is much easier to carry around than a set of scales.

JOHN BEECHAM

From that advice on estimating bear size, it's easy to see that although black bears are all members of the same species, they don't all have the same size, appearance, and physiology. Genetic, environmental, and social factors influence the way bears look, grow, and reproduce (1–5). During our studies, we examined the relationships among nutrition, growth, and sexual maturity in black bears at Council, Lowell, Coeur d'Alene River, and Priest Lake. We also documented differences in coat color and collected blood samples from captured bears at Council and Lowell.

Bears, like other animals and humans, are affected by diseases, parasites, and environmental toxins. These problems can range from mildly annoying to life-threatening. We tested blood samples from some of the bears we captured for selected diseases and environmental toxins; we also checked the bears for external parasites.

Table 5–1. Estimated live weight of black bears for a given neck circumference

Neck Circumference (inches)	Estimated Bear Weight[1] (lb.)	
	Male[2] ($\pm$ 34 lbs.)[4]	Female[3] ($\pm$ 25 lb.)[5]
10	44	39
11	52	47
12	60	56
13	69	64
14	79	74
15	89	84
16	100	95
17	111	106
18	123	118
19	135	131
20	148	144
21	161	158
22	175	—[6]
23	189	—
24	204	—
25	219	—
26	235	—
27	251	—
28	268	—
29	285	—
30	302	—

[1] Based on regression formulas calculated at the 90 percent bound.

[2] Regression formula: Male weight (lb.) $= 0.7588 \times$ (neck circumference in inches)$^{1.7605}$

[3] Regression formula: Female weight (lb.) $= 0.5308 \times$ (neck circumference in inches)$^{1.8709}$

[4] To calculate the weight range for a male, add 34 pounds to the estimated weight and subtract 34 pounds from the estimated weight. For example, a male with a 27-inch neck circumference would weigh between 217 and 285 pounds ($251 - 34 = 217$ and $251 + 34 = 285$).

[5] To calculate the weight range for a female, add 25 pounds to the estimated weight and subtract 25 pounds from the estimated weight. For example, a female with a 17-inch neck circumference would weigh between 81 and 131 pounds ($106 - 25 = 81$ and $106 + 25 = 131$).

[6] No females were captured with neck circumferences exceeding 21 inches.

GENERAL DESCRIPTION

Black bears are medium-sized mammals with relatively short tails. Although adult males occasionally attain fall body weights in excess of 700 pounds, Idaho black bears rarely exceed 400 pounds. Adult females are about 40 percent smaller than males. Young are born in the den in late January or early February and weigh 8 to 12 ounces. Bears have five toes and are *plantigrade*. Their claws, which aren't retractable, are well developed and are usually longer on the forefeet. Their fur is uniform in color, except that they occasionally have a brown muzzle or white markings on the chest. Their dentition is adapted to a diet of plant and animal foods (*bunodont*). Black bears have 42 teeth; their dental formula is 3/3 1/1 4/4 2/3 (Fig. 5–1).

SIZE, GROWTH, AND REPRODUCTIVE PATTERNS

On average, adult male black bears in Idaho were 77 pounds (59 percent) heavier and 6 inches (11 percent) longer than adult females (Table 5–2). Kingsley et al. (6) reported that the spring weights of adult male brown bears were about twice that of females and that differences between sexes in total length were less. Rausch (1) found that female black bear skulls in Alaska were 8 to 11 percent smaller than male skulls after the fifth year. There were no size differences in the first four years.

To determine growth patterns, we used growth curves, which show the relationship between age and weight in a population over a given time period (7). Growth in black bears is *curvilinear*, and is characterized by an initial phase of rapid growth until puberty, followed by a series of seasonal weight gains and losses as individuals approach their maximum size (Fig. 5–2). Seasonal variability in weight is caused by the annual cycle of fat deposition and metabolism. Bears gain weight in late summer/fall and lose body fat during hibernation and the following spring.

We found considerable variation (15 to 30 percent) in the weights of individuals who were the same age (Tables 5–3 through 5–10). Much of this variation was probably caused by genetic differ-

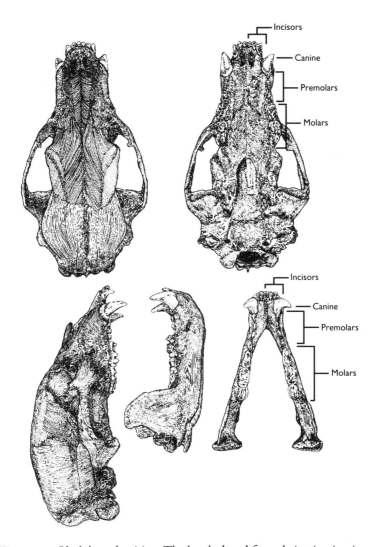

Figure 5–1. Black bear dentition. The bear's dental formula is 3/3 1/1 4/4 2/3 (3 incisors upper jaw/3 incisors lower jaw; 1 canine upper jaw/1 canine lower jaw; 4 premolars upper jaw/4 premolars lower jaw; and 2 molars upper jaw/3 molars lower jaw).

Table 5-2. Weight and total length differences between adult male and female black bears in Idaho

	Weight (lb.)					Total Length (in.)			
Location	N[a]	Males	N	Females	Ratio[b]	N	Males	N	Females	Ratio
Council	5	238	14	142	1.68	5	59	14	52	1.13
Lowell	16	227	27	123	1.85	31	61	37	51	1.20
CDA River	11	206	7	130	1.58	11	59	7	53	1.11
Priest Lake	40	160	12	130	1.23	44	56	12	54	1.04

[a] N = number of bears studied.
[b] Ratio of males to females.

ences, the reproductive status of adult females, year-to-year differences in the quantity and quality of foods available to bears, and small sample sizes. In all the populations we studied, genetic differences were apparent: We observed that physically small females consistently produced smaller cubs than did larger females.

We also saw weight differences based on the females' reproductive status. The demands of lactation resulted in large fluctuations in female weights from one year to the next. Several females with newborn young lost as much as 21 pounds during the spring and summer, and other nursing females gained only a few pounds during the same period. In contrast, females without young gained as much as 50 pounds over the summer.

Another possible cause of the annual variation we observed in adult female weights was the higher metabolic rates of denned females with cubs. In March, the average rectal temperature of denned females accompanied by cubs or yearlings was 98.6°F, about two degrees higher than that of males (96.8°F).

Kingsley et al. (6) reported that older female grizzly bears gained and lost about 70 percent of their spring weight annually; males cycled about 25 percent. Rogers (8) found that some female black bears in Minnesota lost nearly 50 percent of their weight during the reproductive cycle.

Still another cause of weight differences was the availability of foods. Recent research on the food habits, habitat use patterns, and

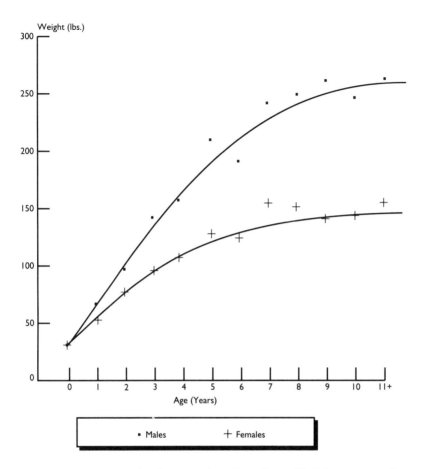

Figure 5–2. Average weights for 223 male and 174 female black bears captured at Council, 1973–77.

social organization of black bears has shown a relationship between nutritional levels and the growth rates and productivity of bear populations (8–13). Although specific growth patterns are due to the evolutionary forces acting on a species, we observed some growth differences within and between the bear populations we studied in Idaho.

The average summer weights of adult males ranged from 250 to 280 pounds and varied significantly only between the Council and

Table 5–3. Mean weight and total length by age class for male black bears captured at Council

Age Class (years)	Weight (lb.)		Total Length (in.)	
	Number Measured	Mean + SD[a]	Number Measured	Mean + SD
0.5	12	31 ± 11.7	10	30 ± 3.4
1.5[b]	28	65 ± 13.6	27	40 ± 2.7
2.5	48	96 ± 24.5	39	46 ± 4.2
3.5	32	141 ± 33.5	25	52 ± 4.3
4.5	20	157 ± 35.7	18	54 ± 3.5
5.5	16	210 ± 49.1	14	57 ± 5.0
6.5	4	188 ± 57.4	3	57 ± 1.2
7.5	9	240 ± 44.8	6	60 ± 2.9
8.5	3	248 ± 51.1	3	57 ± 3.3
9.5	3	262 ± 17.6	3	60 ± 1.8
10.5	1	245	1	53
11–15	3	264 ± 16.9	4	61 ± 2.3
16–20		−		−
20+		−		−

[a] Standard deviation.
[b] Significantly different than 1.5-year-old females.

Priest Lake populations. The average summer weights of adult females ranged from 123 to 141 pounds, and didn't differ significantly except between the Council and Lowell populations. We saw significant differences in weight and total length between 1.5-year-old males and females in the Council and Lowell bear populations (Tables 5–3 through 5–6). These differences also occurred in the Priest Lake population, but not until the bears were 4.5 years old (Tables 5–7, 5–8). In addition, although growth rates were essentially the same for all the populations we studied, bears from Council grew larger than did bears from the other study areas (Tables 5–3, 5–4). These data suggested that nutritional differences rather than genetic influences caused the heavier weights.

In some cases, bear densities can influence bear weights. Because we found similar densities of bears on our study areas, however, we don't believe that density-dependent factors were impor-

Table 5–4. Mean weight and total length by age class for female black bears captured at Council

Age Class (years)	Weight (lb.) Number Measured	Mean + SD[a]	Total Length (in.) Number Measured	Mean + SD
0.5	5	30 ± 9.6	3	29 ± 4.9
1.5[b]	20	52 ± 14.9	19	38 ± 2.1
2.5	20	77 ± 16.5	20	44 ± 3.2
3.5	18	96 ± 16.7	16	46 ± 3.9
4.5	8	108 ± 15.1	7	49 ± 3.8
5.5	15	126 ± 21.9	13	50 ± 2.8
6.5	8	124 ± 20.7	8	51 ± 1.7
7.5	4	156 ± 17.3	4	52 ± 0.3
8.5	3	151 ± 34.1	3	52 ± 2.3
9.5	7	142 ± 27.7	7	52 ± 3.7
10.5	4	144 ± 14.4	4	51 ± 2.5
11–15	6	156 ± 8.7	6	51 ± 2.2
16–20	5	150 ± 13.2	5	51 ± 2.6
20+	4	166 ± 15.0	4	53 ± 1.2

[a] Standard deviation.
[b] Significantly different than 1.5-year-old males.

tant influences on the bears' weight gain.

Population age structures can also affect bear weights: Older, larger bears may control the food resources in an area. We didn't know how population age structures affected the feeding behavior of *sub-adults* in our study areas. Rogers (11) suggested that social factors appeared to regulate which bears, rather than how many bears, could use the food resources in a given area. However, he was unable to document that territorial individuals controlled surplus foods or that they prevented hungry bears from eating them. Egbert and Stokes (14) found that the intensity of aggressive behavior among brown bears in Alaska was strongly related to salmon abundance and they concluded that the availability of food determined, in large part, the social organization of bear populations. We agree with Watson and Moss (15) that both nutrition and behavior influence bear populations and that social organization was probably

Table 5–5. Mean weight and total length by age class for male black bears captured at Lowell

Age Class (years)	Weight (lb.)		Total Length (in.)	
	Number Measured	Mean + SD[a]	Number Measured	Mean + SD
0.5	4	11 ± 3.8	5	22 ± 3.5
1.5[b]	3	50 ± 7.0	5	36 ± 3.9
2.5	11	77 ± 18.1	13	45 ± 3.4
3.5	5	102 ± 11.8	5	50 ± 2.6
4.5	5	112 ± 27.5	6	50 ± 2.7
5.5	9	145 ± 34.7	12	55 ± 4.2
6.5	5	163 ± 24.7	6	56 ± 3.1
7.5	5	225 ± 56.9	6	60 ± 3.5
8.5	3	227 ± 16.5	4	60 ± 1.3
9.5	3	227 ± 38.1	5	62 ± 4.3
10.5	1	230	4	60 ± 2.5
11–15	7	228 ± 36.7	12	60 ± 3.1
16–20	2	225 ± 4.2	6	62 ± 2.6
20+		—		—

[a] Standard deviation.
[b] Significantly different than 1.5-year-old females.

dictated by the availability and distribution of nutritious food supplies.

The quantity and quality of available foods influenced not only bear weights but also their reproductive potential because food resources affected the bears' *age of first reproduction,* litter size, and *litter frequency.* Our analyses of female growth curves in Idaho showed that bears grew rapidly until they reached sexual maturity, and then grew more slowly (Fig. 5–2). We believed that this dramatic change in growth rates was associated with the bears reaching sexual maturity. Wood et al. (16) and Kingsley et al. (6) reported similar relationships between growth rates and puberty in mule deer and brown bears.

Females don't become sexually mature until they reach a certain body size. Because reproduction puts much physiological stress on females, it's advantageous for them to delay breeding un-

Sixty-four percent of the bears we captured had white chest markings. Photo by John J. Beecham

til they are sufficiently large and strong enough to withstand that stress. The similarity in weights of Idaho bears at puberty (Fig. 5–2) suggested that they needed to reach a body weight of about 100 pounds before they produced young. This often occurred when the bears were four to seven years old. Robinette et al. (17) found that mule deer usually weighed at least 90 pounds before they successfully bred. Sadleir (3) reported that reduced availability of food during the juvenile period delayed puberty until an animal attained the minimum body size needed to successfully breed.

Table 5–6. Mean weight and total length by age class for female black bears captured at Lowell

Age Class (years)	Weight (lb.)		Total Length (in.)	
	Number Measured	Mean + SD[a]	Number Measured	Mean + SD
0.5	3	7 ± 1.6	3	19 ± 2.1
1.5[b]	2	37 ± 8.4	2	34 ± 1.4
2.5	5	60 ± 13.5	5	44 ± 5.6
3.5	5	81 ± 12.5	6	49 ± 4.9
4.5	4	98 ± 22.9	6	48 ± 4.4
5.5	8	105 ± 12.3	9	51 ± 3.5
6.5	4	119 ± 7.8	4	50 ± 1.5
7.5	5	99 ± 14.7	5	49 ± 3.3
8.5	7	129 ± 12.0	8	52 ± 1.8
9.5	2	133 ± 7.8	3	53 ± 0.4
10.5	4	114 ± 10.3	5	51 ± 1.4
11–15	13	119 ± 14.0	20	51 ± 2.8
16–20	1	156	1	54
20+		–		–

[a] Standard deviation.
[b] Significantly different than 1.5-year-old males.

The availability of food had other effects on bear reproduction. In the eastern United States, where black bears had access to nutrient-rich nut crops, females produced larger litters, had lower ages of first reproduction and larger average adult body weights (18–21) than in the western United States, where bears fed on a less-rich diet of grasses, forbs, and berries (22, 23, 13). Rogers (8) found that poorly nourished female black bears produced smaller cubs, while captive bears feeding on nutrient-rich diets developed more rapidly than wild bears and commonly matured at 2.5 years of age. Sauer (24) reported that female bears in New York reached adult size by 2.5 years and were sexually mature. Sadleir (3) concluded that most of the difference in age at puberty in wild animals was probably due to differences in growth rates.

After analyzing growth characteristics for black bears in Idaho, we concluded that none of the populations we studied

Table 5–7. Mean weight and total length by age class for male black bears captured at Priest Lake

Age Class (years)	Weight (lb.)		Total Length (in.)	
	Number Measured	Mean + SD[a]	Number Measured	Mean + SD
0.5	8	20 ± 7.8	8	27 ± 4.1
1.5	10	57 ± 19.7	9	39 ± 3.7
2.5	23	81 ± 19.0	20	45 ± 3.2
3.5	30	115 ± 32.2	20	50 ± 4.5
4.5[b]	21	142 ± 32.8	14	52 ± 4.5
5.5	13	169 ± 33.0	8	55 ± 4.1
6.5	13	205 ± 45.9	11	59 ± 4.3
7.5	5	197 ± 43.9	4	56 ± 2.9
8.5	5	195 ± 30.1	5	57 ± 4.3
9.5	4	220 ± 30.0	2	60 ± 0.8
10.5	3	226 ± 6.9	4	63 ± 2.9
11–15	5	195 ± 22.1	5	56 ± 3.1
16–20	1	230	–	–
20 +	1	205	1	62

[a] Standard deviation.
[b] Significantly different than 4.5-year-old females.

achieved their maximum growth or reproductive potential. Habitat quality appeared to be the primary factor limiting growth rates, but behavior- or density-related elements were possible limiting factors. Nutrition was a primary factor in litter size, litter frequency, and the age of first reproduction in bear populations (13). We don't know the impact of behavioral differences on the age of first reproduction in bear populations with different densities or age structures.

We also concluded that the age of first reproduction may be an indicator of habitat quality. Because of the strong relationship between nutrition and the productivity of bear populations, we believed that major vegetative changes would affect growth rates, the age of first reproduction and, therefore, the reproductive potential of the population. Habitat manipulation may be a viable management tool for increasing or enhancing the quantity and quality of

Table 5–8. Mean weight and total length by age class for female black bears captured at Priest Lake

Age Class (years)	Weight (lb.)		Total Length (in.)	
	Number Measured	Mean + SD[a]	Number Measured	Mean + SD
0.5	2	19 ± 1.4	2	27 ± 0.5
1.5	5	48 ± 24.8	5	37 ± 3.7
2.5	2	70 ± 17.0	2	42 ± 2.9
3.5	6	94 ± 24.8	6	47 ± 2.6
4.5[b]	10	105 ± 17.8	9	49 ± 4.0
5.5	13	117 ± 16.0	12	50 ± 2.5
6.5	10	124 ± 18.2	10	51 ± 2.6
7.5	4	127 ± 7.8	4	51 ± 2.1
8.5	3	119 ± 4.9	3	50 ± 5.3
9.5	–	–	–	–
10.5	4	133 ± 8.9	4	53 ± 1.8
11–15	9	133 ± 18.3	9	54 ± 3.6
16–20	–	–	–	–
20+	2	140 ± 0	2	51 ± 1.2

[a] Standard deviation.
[b] Significantly different than 4.5-year-old males.

bear habitat, thus improving the productivity of Idaho bear populations.

COLOR PHASES

The black bear is one of several species of mammals in North America who have distinct color phases. Throughout most of eastern North America, where early explorers first saw and described black bears, bears are predominantly black in color. However, in the western half of the continent, the brown color phase is common and in some areas occurs more frequently than the black phase. Two other color phases are also found along the west coast of Canada and Alaska. The white (Kermode) and blue (Glacier) color phases are found in west-central British Columbia and in southeast Alaska, respectively. The Kermode black bear has a white color phase with brown eyes and reddish foot pads. It isn't an albino, although albi-

Table 5–9. Mean weight and total length by age class for male black bears captured at Coeur d'Alene River

Age Class (years)	Weight (lb.) Number Measured	Mean + SDᵃ	Total Length (in.) Number Measured	Mean + SD
0.5	3	25 ± 18.0	4	32 ± 5.8
1.5	5	41 ± 9.9	5	36 ± 2.1
2.5	11	97 ± 16.4	11	49 ± 2.8
3.5	2	151 ± 13.4	2	58 ± 1.1
4.5	1	130	1	54
5.5	–	–	–	–
6.5	5	199 ± 34.9	6	60 ± 4.3
7.5	1	190	1	59
8.5	4	191 ± 27.6	4	59 ± 1.6
9.5	5	216 ± 40.7	5	63 ± 3.7
10.5	3	203 ± 32.1	4	58 ± 3.6
11–15	6	223 ± 22.3	6	1 ± 3.5
16–20	2	197 ± 24.8	2	60 ± 2.0
20+	–	–	–	–

ᵃ Standard deviation.

nism has been reported in the literature (25). The Glacier or blue bear varies in color from silver-gray to a gun-barrel bluish gray.

Cowan (26) first described the distribution of black bear color phases based on Hudson's Bay Company fur trading records. He didn't, however, attempt to explain why bears exhibit this characteristic or how it is maintained in their populations. Gershenson (27) presented considerable evidence that natural selection was the most plausible explanation for the occurrence of different color phases in hamsters in the Soviet Union. Jonkel (28) carried this idea a step further by suggesting that brown-phase bears experienced lower mortality rates than black-phase bears and that brown-phase bears made more efficient use of their habitat. He wasn't able to demonstrate differences in mortality rates between the two color phases, but he did document that brown-phase bears fed longer into the day in the more open, productive feeding areas.

Although no specific studies have been done to examine the

Table 5–10. Mean weight and total length by age class for female black bears captured at Coeur d'Alene River

Age Class (years)	Weight (lb.)		Total Length (in.)	
	Number Measured	Mean + SD[a]	Number Measured	Mean + SD
0.5	3	25 ± 8.7	3	28 ± 2.4
1.5	–	–	–	–
2.5	5	70 ± 29.8	5	47 ± 2.4
3.5	3	76 ± 7.1	3	44 ± 2.1
4.5	1	132	1	50
5.5	–	–	–	–
6.5	1	100	1	55
7.5	5	119 ± 15.2	5	51 ± 2.8
8.5	1	120	1	51
9.5	1	124	1	54
10.5	1	148	1	57
11–15	6	124 ± 29.4	6	51 ± 3.3
16–20	–	–	–	–
20+	–	–	–	–

[a] Standard deviation.

genetic relationships among these color phases, most scientists now agree that color phases are genetically controlled and that their distribution is governed by weather conditions that favor one color phase over another through the process of natural selection. In the ecological literature, Gloger's rule states that among warm-blooded animals, black pigments are most prevalent in warm, humid areas, and reds and yellows are more common in dry areas. The distribution of black bear color phases follows this rule closely throughout the bears' range. Brown-phase bears are found more often in hot, dry climates, while black-phase bears are more common in humid areas that receive lesser amounts of solar radiation. A. LeCount and T. Waddell reported that 59 percent and 94 percent of the bears they captured in two Arizona studies were brown-phase bears and T. Beck reported that 95 percent of the bears he handled in Colorado were brown (29).

We observed significant differences in the distribution of black

Table 5–11. Distribution of black and brown color phases and white chest markings of black bears in Idaho

	Color Phase		Whitespot	
Area	Black	Brown	Yes	No
Council	48	148	145	51
Lowell	86	71	106	51
CDA River	39	23	24	29
Priest Lake	93	60	81	72

and brown color phases in Idaho (Table 5–11). The bears we captured at Council were predominantly brown (76 percent), while the majority of bears we captured in northern Idaho were black (range 54 to 63 percent).

We believe that the distribution of color phases in Idaho black bears is related to the state's weather patterns. Weather records indicate that northern Idaho receives less solar radiation (55 percent of the total sunshine possible) than southern Idaho (66 percent) and has lower evaporation rates. The ability of brown-phase bears to tolerate heat (28) probably gives them a selective advantage over black-phase bears in the open, arid habitats of southern Idaho, but gives them little or no advantage in those areas of northern Idaho with low or moderate levels of solar radiation.

Although we didn't see any bears change from brown-phase to black-phase bears or vice versa, we did observe changes in color among brown-phase bears (Plate 2). Straw or blond bears shed their winter/spring coats in the summer and grew new coats that were a darker, chocolate brown. Several bears we captured were light brown as subadults and dark brown as adults. This tendency to turn dark brown as an adult was particularly evident in males; we captured fewer than five blond adult males. However, it wasn't uncommon to capture adult blond females in the spring. We saw no difference in the proportion of black and brown color phases between subadult and adult bears, and weren't able to document any difference in mortality patterns between black- or brown-phase bears in Idaho.

Table 5–12. Color phases of black bear litters in Idaho

	Litters			
Female Color	Number Studied	Black	Brown	Mixed
Black	39	27	6	6
Brown	33	5	20	8
Total	72	32	26	14

Family groups (females with cubs) usually had the same color phase, although we did observe mixed-color-phase litters (Plate 3). Black-phase females produced predominantly black-phase litters (69 percent); brown-phase females produced mostly brown-phase litters (61 percent). Fourteen of 72 litters (19 percent) were mixed-color-phase litters (Table 5–12).

The distribution of bears with white chest markings was variable. Although a majority of bears had white spots at Council, Lowell, and Priest Lake, only bears captured at Council and Lowell had a significantly higher proportion of these markings. We saw no differences in the number of black- or brown-phase bears who had white spots on their chests or in the number of males and females with white chest markings.

DISEASES

We took 352 blood samples from 265 bears captured at Council and Lowell to conduct blood serum surveys for selected infectious diseases. We found medium to high *titers* for tularemia, brucellosis, and toxoplasmosis in some bears. We observed low titers for leptospirosis, Q fever, St. Louis encephalitis, western equine encephalitis, and Rocky Mountain spotted fever. Ten of the bears sampled had titers to more than one disease at a single sampling.

Prior to these studies, researchers hadn't reported tularemia or brucellosis antibodies in black bears. *Toxoplasma* had been reported in black bears from Ontario, Canada, but not in bears from the United States (30, 31).

We saw no evidence of clinical disease. In the case of tularemia, we observed large fluctuations in *antibody* titers, which suggested that some bears were capable of surviving significant exposure to the disease.

We don't know if the species of *Brucella* found in Idaho black bears produces a positive titer for *B. abortus,* a disease that causes cattle infections. This relationship should be further studied.

PARASITES

Although we didn't conduct a systematic survey of internal or external parasites of Idaho black bears, we did routinely record the presence of external parasites on captured bears. Ticks were by far the most common external parasite found on black bears captured in Idaho. Although ticks were associated with at least one disease (tularemia) in black bears, we found that most bears were lightly infested (had fewer than 25 ticks). We found lice, fleas, and mites infrequently on black bears but didn't consider them important parasites.

Four bears we captured showed clinical signs of mange, a skin condition associated with mites, but only one bear had mites, which were found in deep skin scrapings. This bear, when originally captured at Lowell in July 1975, had *chronic lesions* on the muzzle and forehead and *acute lesions* on the neck, chest, and front legs (32). Thirteen months later, we recaptured this same bear, and 50 to 70 percent of the hair on the muzzle and forehead had regrown to its normal length and thickness. Although the skin had become melanotic (darker), all the lesions we observed in 1975 had improved. The species of mite we found on this bear (*Ursicoptes americanus*) was known only from the original description of three female specimens taken from a black bear that died in a Kansas zoo (33). We found, for the first time, the male nymph and larva of this species, which aided in resolving the questionable taxonomic status of this family of mites.

The incidence of trichinosis (an internal parasite) we observed in Idaho black bears – 13 percent – was similar to that reported for black bears studied in Glacier and Yellowstone national parks (34).

However, it was higher than the 2.3 percent reported by Zimmerman (35) for Idaho black bears and for black bears in other parts of the United States (34).

PESTICIDE AND MERCURY LEVELS

The average residue levels for both pesticides and mercury found in bears in this study were well below the maximum tolerances allowed in domestic livestock by the U.S. Food and Drug Administration and weren't considered important from a population perspective.

LITERATURE CITED

1. RAUSCH, R. L. 1961.
Notes on the black bear (*Ursus americanus* Passas) in Alaska, with particular reference to dentition and growth. Z.f. Saugetierkunde. 26:65–128.
2. RICKLEFS, R. E. 1968.
Patterns of growth in birds. Ibis. 110:419–51.
3. SADLEIR, R. M. F. S. 1969.
The role of nutrition in the reproduction of wild mammals. J. Reprod. Fer. 6:39–48.
4. WEINER, J. G., AND S. E. FULLER. 1975.
Comments on "Environmental evaluation based on relative growth rates of fishes." Prog. Fish Culturist. 37(2):99–100.
5. FENDLEY, T. T., AND I. L. BRISBIN, JR. 1977.
Growth curve analyses: A potential measure of the effects of environmental stress upon wildlife populations. XIII Congr. Game Biol. 337–50.
6. KINGSLEY, M. C. S., J. A. NAGY, AND R. H. RUSSELL. 1983.
Patterns of weight gain and loss for grizzly bears in northern Canada. Int. Conf. Bear Res. and Manage. 5:174–78.
7. KINGSLEY, M. C. S. 1979.
Fitting the von Bertalanffy growth equation to polar bear age-weight data. Can. J. Zool. 57:1020–25.
8. ROGERS, L. L. 1976.
Effects of mast and berry crop failures on survival, growth, and reproductive success for black bears. Proc. North Am. Wildl. and Nat.

Resour. Conf. 41:431–38.

9. BEEMAN, L. E. 1975.
Population characteristics, movements, and activities of the black bear (*Ursus americanus*) in the Great Smoky Mountains National Park. Ph.D. thesis, University of Tennessee, Knoxville. 218pp.

10. CHERRY, J. S., AND M. R. PELTON. 1976.
Relationships between body measurements and weight of the black bear. J. Tenn. Acad. Sci. 51(1):32–34.

11. ROGERS, L. L. 1977.
Social relationships, movements, and population dynamics of black bears in northeastern Minnesota. Ph.D. thesis, University of Minnesota, Minneapolis. 194pp.

12. ALT, G. L. 1980.
Rate of growth and size of Pennsylvania black bears. Pa. Game News 51(12):7–17.

13. REYNOLDS, D. G., AND J. J. BEECHAM. 1980.
Home range activities and reproduction of black bears in west central Idaho. Int. Conf. Bear Res. and Manage. 4.

14. EGBERT, A. L., AND A. W. STOKES. 1976.
The social behaviour of brown bears on an Alaskan salmon stream. *In* M. R. Pelton, J. W. Lentfer, and G. E. Folk, eds. Bears – Their biology and management. IUCN New Ser. 40:41–56.

15. WATSON, A., AND R. MOSS. 1971.
Spacing as affected by territorial behavior, habitat, and nutrition in red grouse (*Lagopus lagopus scoticus*). Pages 92–111. *In* A. H. Esser, ed. The use of space by animals and men. Plenum Press, N.Y. 411pp.

16. WOOD, A. J., I. MCT. COWAN, AND H. C. NORDAN. 1962.
Periodicity of growth in ungulates as shown by deer of the genus *Odocileus*. Can. J. Zool. 40:593–603.

17. ROBINETTE, W. L., C. H. BAER, R. E. PILLMORE, AND C. E. KNITTLE. 1973.
Effects of nutritional change on captive mule deer. J. Wildl. Manage. 37:312–26.

18. SPENCER, H. E., JR. 1955.
The black bear and its status in Maine. Maine Dept. Inland Fish and Game Bull. 4. 55pp.

19. ERICKSON, A. W., AND J. E. NELLOR. 1964.
Breeding biology of the black bear. Michigan State Univ. Agric. Exp. Stn. Res. Bull. 4:5–45.

20. HAMILTON, R. 1972.
Summaries, by state: North Carolina. Pages 11–13. *In* R. L. Miller, ed. Proceedings of the 1972 black bear conference. N.Y. State Dept. of Environ. Conserv., Delmar, N.Y.

21. COLLINS, J. M. 1973.
Some aspects of reproduction and age structures in the black bear in North Carolina. Proc. Ann. Conf. S.E. Assoc. Game and Fish Comm. 27:163–70.

22. BARNES, V. G., AND O. E. BRAY. 1967.
Population characteristics and activities of black bears in Yellowstone National Park. Final Rep. Colo. Coop. Wildl. Res. Unit, Fort Collins, 196pp.

23. JONKEL, C. J., AND I. M. COWAN. 1971.
The black bear in the spruce-fir forest. Wildl. Monogr. No. 27. 55pp.

24. SAUER, P. R. 1975.
Relationships of growth characteristics to sex and age for black bears from the Adirondacks region of New York. New York Fish and Game J. 22(2):81–113.

25. STANDLEY, P. C. 1921.
Albinism in the black bear. Sci. 54 (1386):74.

26. COWAN, I. MCT. 1938.
Geographic distribution of color phases of the red fox and black bear in the Pacific Northwest. J. Marcum. 19(2):202–6.

27. GERSHENSON, S. 1945.
Evolutionary studies on the distribution and dynamics of melanism in the hamster (*Cricetus cricetus* L.). I. Distribution of black hamsters in the Ukrainian and Bashkirian Soviet Socialist Republics (U.S.S.R.). Genetics. 30:207–32.

28. JONKEL, C. J. 1967.
The ecology, population dynamics, and management of the black bear in the spruce-fir forests of northwestern Montana. Ph.D dissertation, University of British Columbia, Vancouver.

29. LECOUNT, A., ARIZ.
Game and Fish Dept., Phoenix, T. Waddell, Ariz. Game and Fish Dept., Phoenix, and T. Beck, Colo. Div. of Wildlife, pers. commun.

30. QUINN, P. J., R. O. RAMSDEN, AND D. H. JOHNSTON. 1976.
Toxoplasmosis: A serological survey in Ontario wildlife. J. Wildl. Dis. 12:504–10.

31. TIZARD, I. R., J. B. BILLETT, AND R. O. RAMSDEN. 1976.
The prevalence of antibodies against *Toxoplasma gondii* in some Ontario mammals. J. Wildl. Dis. 12:322–25.

32. YUNKER, C. E., C. E. BINNINGER, J. E. KEIRANS, J. J. BEECHAM, AND M. SCHLEGEL.
Clinical mange in the black bear associated with *Ursicoptes americanus* (*Acari: Audycoptidae*). J. Wildl. Dis. 16(3):347–56.

33. FAIN, A., AND D. E. JOHNSTON. 1970.
Un nouvel acarien de la famille Audycoptidae chez l'ours noir *Ursus Americanus* (*sarcoptiformes*). Acta Zool. Pathol. Antverpiensia. No. 50:179–81.

34. WORLEY, D. E., J. C. FOX, J. B. WINTERS, AND K. R. GREER. 1974.
Prevalence and distribution of *Trichinella spiralis* in carnivorous mammals in the U.S. Northern Rocky Mountain Region. *In* C. W. Kim, ed. Trichinellosis, Proc. Third International Conference on Trichinellosis. Intext Educational Publishers, New York, N.Y.

35. ZIMMERMAN, W. J. 1977.
Trichinosis in bears of western and north-central United States. Am. J. Epidemiol. 106:167–71.

SIX | BEHAVIOR AND SOCIAL ORGANIZATION

IN ANY ANIMAL CAPTURE STUDY, YOU ALWAYS EN-
counter individual animals who are either trap-shy (very difficult to trap) or trap-prone (you catch them more often than
you would like). At Priest Lake, one radio-collared female,
No. U-680, often bedded within 50 yards of a trap site and fed
in the immediate vicinity of the site without attempting to eat
the bait we placed in the "cubby." She was exceptionally
wary, and we caught her very few times during the study. It
was enough to make us feel personally rejected.

Another Priest Lake bear, No. U-269, showed quite the
opposite behavior. This adult male was trapped nine times in
1979; no other bear was caught so frequently during our studies. No. U-269 appeared to know he was going to get snared,
but just didn't seem to care. Unlike most bears we trapped
more than once, No. U-269 didn't try to avoid the snare or
destroy the cubby to get at our bait. In fact, several times we
found him at a trap site with the snare hanging loosely around
his leg, the drag log unmoved and all our bait gone. He was
patiently waiting for us to show up and turn him loose. We finally decided he had had a deprived cubhood and needed our
constant attention to compensate!

Turning No. U-269 loose, however, was no easy task. He
wasn't restricted by the drag log and always seemed intent on

maintaining his dignity by repeatedly charging us and vocalizing before we could get close enough (within 8 feet) to inject
the tranquilizing drugs into him. On one occasion, Jeff spent
over four hours unsuccessfully trying to tranquilize No. U-
269. We eventually got the job done, but it took all our efforts
and those of several campers who happened to pass by.

JOHN BEECHAM

When are bears most active? Where do they travel during the year?
Do they defend their home ranges? How do bears behave when
they're threatened? Before sophisticated radio-telemetry equipment was available, it was hard to answer these questions. Now the
widespread use of radio-telemetry equipment has allowed wildlife
managers to collect information on black bear activity patterns,
movements, home range characteristics, and behavior from many
areas in North America. These data are essential to understanding
black bear population dynamics and they aptly illustrate the adaptive nature of black bears (1–10).

During our studies at Council and Priest Lake, we obtained
3,780 radio locations on 54 bears (26 males; 28 females). We collected 3,096 locations from 46 bears at Council during 1973–1976
and 1982–1983, and 684 locations from 8 bears at Priest Lake in
1980–1981.

We studied activity patterns and daily and seasonal movements. We moved some bears to new areas and monitored their
movements to see if they returned home (homing behavior). We
also examined several characteristics of the bears' home ranges. We
looked at home range use, fidelity (whether they use the same
home range each season or year), and overlap (whether they allow
other bears on their home range). Finally we watched the bears we
captured to see how they reacted to our presence.

ACTIVITY PATTERNS

Black bears were originally thought to be primarily nocturnal animals because they were often seen at night around campgrounds or

summer homes. However, our telemetry studies clearly showed that these activity patterns were an adaptation to the presence of humans and that wild bears in a natural setting are seldom active at night. At Council and Priest Lake, black bears were significantly more active during daylight hours. Generally, bears showed major activity peaks between 5:00 a.m. and 10:00 a.m. and between 6:00 p.m. and 9:00 p.m. They were least active between 1:00 a.m. and 4:00 a.m. Bears at Council were less active and more nocturnal in early spring (March-May) and late fall (November) than in summer and early fall (June-October, see Table 6–1). Similar activity patterns have been reported for black bears in other areas (1, 3, 7).

Table 6–1. Effect of season on activity patterns of nine radio-collared bears at Council in 1974

Time Period	% of time bears were active*		
	Sunrise-Sunset	Night	Total
March–May; November	32	75	48
June–October	79	37	64

* Based on 528 radio locations.

Our analysis of monthly activity data suggested that bear activity levels changed during the year. Radio-collared bears generally left their den sites immediately after emerging in the spring, but because food supplies were of marginal quality, the bears weren't fully active. Their activity levels increased gradually as spring progressed and foods became more available. In the summer, when energy-rich nutritious foods were available, the bears were most active. Activity levels declined in the fall, and some bears became quite lethargic when berry crops were no longer available. On September 25, 1977, one of our students got within 20 yards of an inattentive radio-collared female (No. U-41). We observed this type of behavior on several occasions with females before they denned. Males tended to den immediately after arriving at their den sites, so we rarely saw any lethargic behavior in them.

MOVEMENTS

In their search for food, bears moved up and down in elevation. In early spring, we found black bears at lower elevations, feeding on newly emerged grasses and broadleaf forbs. As a rule, the bears followed the "green-up" up mountain slopes as winter snows melted. By mid-July, when the snows had melted and the grasses and forbs at lower elevations were no longer palatable, the bears tended to feed at higher elevations. They moved to middle and lower elevations in late July to eat huckleberries and buffaloberries, but returned to higher elevations in August. In September, Council bears again moved to lower elevations to feed on chokecherries and hawthorn. Bears at Priest Lake showed similar movements in September, traveling to lower elevations to feed on bearberries and devil's club berries. Bears at Council and Priest Lake generally remained at lower elevations until denning.

The patchy distribution of food influenced the bears' day-to-day movements within their home ranges at Council. Bears didn't use established trails, but did use the same travel routes to move from one area of their range to another. Herrnstein (11) found that domestic pigeons learned to peck at disks in direct proportion to the number of times each disk rewarded the pigeon with food. We believed that as bears moved among food patches at Council, they probably had varying success in obtaining food. When they reached their home range boundary or an area where food was scarce, they retraced the route along which they had found the most food. We considered these movement patterns to be an efficient feeding strategy for exploiting patchily distributed food resources (6).

Black bears moved within their home ranges to take advantage of locally abundant foods, but also traveled to areas of plentiful foods outside their traditional home ranges. Two Council bears moved 7 and 12 miles, respectively, from the center of their home ranges, where food was scarce, to lower elevations where berries were abundant. In August 1975, we captured an unmarked 16-year-old male (No. U-118), who remained on our Council study area for two months and then disappeared. He was killed in April 1977 on the Middle Fork of the Payette River, about 32 miles southeast of where we tagged him. We believed this bear lived in the Payette River drainage and was captured at Council while making an excursion outside his traditional home range. We observed similar extended movements by ear-tagged bears during drought years at Priest Lake.

We found that adult males were more mobile than females during all months. The mean distances traveled by bears between radio locations remained relatively constant from month to month, suggesting that male black bears didn't travel more during the breeding season, as reported by Lindzey (12).

We saw no difference in movements between females with and without cubs. The home range sizes for 7 of 12 Council females accompanied by cubs were similar. Four females had larger home ranges in years when they were with cubs and one had a smaller home range. These data indicated that the presence of cubs doesn't restrict the movements of lactating females, who remain quite mobile.

HOMING MOVEMENTS

Black bears who kill livestock or live near human population centers are often captured and moved to areas where they will cause fewer problems for people. However, sometimes the bears will return to their home ranges. The success of bear relocation efforts varies considerably and is probably affected by the sex and age of the bear, the availability of natural foods and the bear's prior experience with humans.

During the 1976 field season at Lowell, we captured and moved 72 black bears (38 males; 34 females) as part of a department-sponsored calf elk mortality study. Three other bears were captured in 1976 but weren't released: One had a severe case of mange and was put to sleep; another had been shot and severely wounded by a hunter, so was also put to sleep; and a cub was killed when his mother stepped on him in a holding cage. We ear-tagged and released the other 72 bears in 12 locations (Fig. 6–1) scattered

Figure 6–1. Release sites for 72 bears captured at Lowell in 1976. Lines show the distance two females and five males traveled to return to Lowell.

throughout the northern two-thirds of Idaho. In 1977 and 1979, we recaptured 7 (5 males; 2 females) of the relocated bears on the Lowell study area. These bears had traveled an average distance of 88 miles to return to the Lowell area. The males had gone an average of 86 miles, while the two females had traveled 82 and 109 airline miles to get back to the study area. Several bears showed strong homing instincts and were recaptured at the same trap sites they had been caught in during 1976.

In subsequent years, we recovered an additional 19 bears (14 males; 5 females). We captured two of these bears at their release location at Council during routine trapping efforts in 1976 and 1977. Of the remaining 17 bears, 14 were shot by hunters, 2 were killed by sheepherders, and 1 died from unknown causes in 1980. Two of the recovered bears had traveled back toward Lowell. The other 17 bears showed no indication of homing. In fact, some bears went off in random directions: 3 of 8 Lowell bears released on Lunch Peak, 160 miles northwest of Lowell, traveled away from the Lowell study area. One bear moved north into Canada, another moved southeast into Montana, and the third went south toward Clark Fork, Idaho. Hunters shot two of those bears in October 1976 and one in May 1977.

These data suggested that some individuals, both males and females, have strong homing instincts and can move long distances over extremely rough, mountainous terrain to return to their home ranges. Other bears showed no tendency to home and stayed at their release locations or left them in random directions. Similar movements and homing data were reported by Payne (13), Beeman and Pelton (14), McArthur (15), and Alt (16) for black bears and by Miller and Ballard (17) for brown bears.

COUNCIL HOME RANGE CHARACTERISTICS

Home Range Use

Home range sizes varied, depending on the bears' sex and age, and whether they were residents or transients who were *dispersing*. The 8

We placed radio collars on bears at Priest Lake and Council to learn more about black bear behavior. Photo by John J. Beecham

adult males we studied occupied significantly larger home ranges than did the 33 adult females (Table 6–2). The average annual home range size for adult males was 56 square miles, while that for adult females averaged 12 square miles. Stickley (18), Erickson and Petrides (19), Jonkel and Cowan (20), Poelker and Hartwell (1), and Lindzey (12) reported similar relationships in the size of adult male and female home ranges. Adult males probably inhabited larger home ranges than females to increase their breeding opportunities. Under optimum conditions, adult female bears produced young every other year. Therefore, it was advantageous for males to establish home ranges that encompassed four to six female home ranges to increase their odds of breeding each year (2, 5, 21).

Cubs stayed with their mothers in the mother's home range during their first year of life. The following spring, yearlings remained with their mothers until early June, when the family broke up. We monitored four family groups in 1975 and 1976 to document family breakups. Of the four family groups, two split up between

Table 6–2. Average annual home range size (square miles) for radio-collared black bears at Council and Priest Lake

Location	Age	Males Number Studied	Males Home Range Size	Females Number Studied	Females Home Range Size
Council	Yearlings	7	19 mi² (range 8–47 mi²)	3	5 mi² (range 3–8 mi²)
	Subadults	3	29 mi² (range 13–44 mi²)	1	17 mi² –
	Adults	8	56 mi² (range 12–64 mi²)	33	12 mi² (range 3–35 mi²)
Priest Lake	Adults	5	16 mi² (range 2–40 mi²)	6	5 mi² (range 3–7 mi²)

May 24 and June 3, the third broke up between May 27 and June 3, and the fourth group separated between May 30 and June 4. After the family breakups, there were occasional reunions between the mother and one or more yearlings and between yearlings. In one instance, we saw a mother and her yearling sleeping under a tree during a rainstorm several weeks after the family breakup; the yearling was lying on top of the female.

Following the family breakups, 9 of 10 radio-collared yearlings remained in their mothers' home range and denned there the following fall. The other yearling male left the study area and traveled about 12 miles south, where he used a 2-square-mile area until he denned in November. As the season progressed, the average distance between females and their yearlings and between siblings increased. These data suggested that the family bond became progressively weaker as time passed.

Even though both male and female yearlings remained within their mothers' home range, males used significantly larger areas than females. The average home range size for yearling males was 19 square miles (Table 6–2); for yearling females, it was 5 square miles.

Home range sizes for subadult bears also varied. In our trapping studies at Council during 1975 and 1976, we captured both resident and transient subadults. Of 52 bears less than four years old, we captured 23 (44 percent) more than once. We knew 14 of these 23 bears were residents, but we considered the 29 bears captured only once to be transient bears that were dispersing. Subadult males made up 56 percent of the resident group and 93 percent of the dispersing group.

The size of a subadult bear's home range depends on its sex and social status (whether it is a resident or a dispersing bear). Subadult males at Council used home ranges that averaged 29 square miles in size, which were intermediate in size to the adult males' average range (56 square miles) and the adult females' average range (12 square miles). Subadult females used home ranges that were comparable in size (they averaged 17 square miles) to that of adult females. Eveland (22) reported that subadult males in Pennsylvania used larger home ranges than bears of other sex/age classes, but he may have included dispersing subadults in his sample. Dispersing subadults roamed farther than resident bears.

Home Range Fidelity

We saw no significant differences in the home ranges used by adult females from one year to the next. Adult males, in contrast, occasionally made dynamic shifts in home ranges from year to year. Between 1975 and 1976, when the huckleberry crop improved from very poor to good, three of four adult males shifted home ranges. Their centers of activity differed significantly between years, and 50 to 80 percent of their radio locations in 1976 were outside their previous home ranges. We believe that the availability of foods, particularly huckleberries and serviceberries, was responsible for the dramatic shifts in home range in those years.

Home Range Overlap

Black bears at Council showed no evidence of territorial behavior. They didn't exclude other bears from their home ranges, and on several occasions we observed bears feeding close to one another without displaying aggressive behavior. The minimum home range overlap for radio-collared males ranged from 54 to 100 percent; for females, it was 34 to 89 percent. Home range overlap between the sexes was close to 100 percent. If all bears using the study area had been radio-collared, we believed that home range overlap for both males and females would have been complete.

The Council bear population contained many young animals. We believe that the high degree of tolerance we saw was probably influenced in part by the population's young age structure and the tendency of young bears to be less aggressive. Another factor that undoubtedly contributed to the extensive home range overlap and high tolerance was the distribution of food on the study area. Horn (23) demonstrated that it wasn't advantageous for animals to defend fixed areas with patchy and unpredictable food supplies. Weins (24) agreed with Horn and suggested that the patchiness of resources, such as food, governed the pattern of social and space-related behavior of a population. Where food resources are abundant, evenly distributed, and predictable, territorial behavior may be the best strategy for a species. However, as these resources become unevenly distributed or more unpredictable, bears have to expand their home ranges to get enough food. At some point, bears may expend too much energy defending their areas and territoriality is no longer a profitable way to limit overlap. Then it becomes more advantageous for them to establish home ranges and allow overlap by other bears than to attempt to defend exclusive territories.

Other studies have shown that *intraspecific tolerance* is variable among black and grizzly bears (25, 20, 26, 1, 12). The distribution of food and a population's age structure and social relationships are interrelated factors affecting the social organization and behavior of bears.

We also wanted to see if bears minimized contact with other bears of the same sex, so we outlined the smallest area that would encompass 75 percent of each bear's radio location points. Overlap in these areas was minimal for females but remained high for males. Females appeared to minimize contact with nearby females by concentrating their activities in a portion of their total home range. Lindzey (12) found similar home range overlap for males and fe-

Some of the male black bears we studied were scarred, probably due to fights over female bears during the breeding season. Photo by John J. Beecham

males in Washington and concluded that a dominance hierarchy produced the spatial and temporal separation he observed among females. Reynolds and Beecham (6) suggested that avoidance behavior (females tended to avoid each other) also played a role in establishing spacing patterns.

PRIEST LAKE HOME RANGE CHARACTERISTICS

Home Range Use

The five adult males we studied used larger home ranges in all seasons than did the six females we studied at Priest Lake (Table 6–2). The average annual home range size was 16 square miles for males and 5 square miles for females. Although the ratio of male to female home range size at Priest Lake was similar to the ratio we observed at Council, both males and females occupied significantly smaller areas at Priest Lake, perhaps because the food supply at Priest Lake was more stable.

Home Range Fidelity

Black bears at Priest Lake showed seasonal and annual home range fidelity. The distribution and availability of foods along the west side of the Selkirk Mountains was relatively uniform and predictable. As a result, bears at Priest Lake maintained fairly stable, small home ranges from one year to the next. Reynolds and Beecham (6) reported stable home ranges for adult females at Council, but not for males. Lindzey and Meslow (3), Rogers (4), Alt et al. (5), Garshelis and Pelton (7), and Schwartz et al. (10) also reported home range fidelity in black bears.

Home Range Overlap

We couldn't adequately assess the amount of home range overlap at Priest Lake because few bears were radio-collared. However, home range overlap was common for collared bears, and our field observations of noncollared bears further suggested that overlap was extensive among bears. We also saw barren females (No. U-650 and U-680) feeding near other bears in the fall, suggesting some intraspecific tolerance within individual home ranges.

In spite of this tolerance, we saw wounds and scarring on our trapped male black bears. We believed this was due to competition between males during the breeding season. This same behavior was observed at Lowell and Coeur d'Alene River, but not at Council until 1982, when most bears in the population were older (the median age was 5.5 years).

BEHAVIOR AT TRAP SITE

During 1975–1977, we documented the behavior of black bears captured in leg-hold snares to see how they reacted to us (Plate 4). We recorded observations when we first arrived at the trap site and when we approached the bear to administer the immobilizing drugs. We collected 364 observations from 106 bears captured on 182 occasions. From these observations, we identified six distinct behaviors:

1. *Threat* – This behavioral response varied from low-intensity reactions (moaning or growling) to full charges toward us. Low-intensity reactions were common. High-intensity threats involving full charges or bluff (fake) charges and vocalizations (moaning, huffing, and jaw chomping) occurred less frequently. We often heard vocalizations during threat behaviors, but heard none during full charges. Forty percent of our observations involved low-intensity reactions, 6 percent involved bluff charges and vocalizations, and 3 percent involved full charges.

2. *Redirected* – In this behavioral response, a snared bear redirected its aggression from us to some other object nearby or to itself. We saw this behavior in less than 2 percent of our observations.

3. *Flight* – This behavior occurred when a bear attempted to run from us or climb a tree to escape. The bear often made huffing noises as it tried to flee. Flight behavior accounted for 15 percent of our observations.

4. *Hiding* – This behavior was characterized by a bear covering its head with its paws or forearms or seeking cover in shrubs or behind trees and logs. Hiding behavior occurred in 5 percent of our observations.

5. *Submissive* – Submissive bears usually sat with their heads facing away from us. They frequently moaned, bawled, or made huffing sounds. Submissive reactions made up 15 percent of our observations.

6. *No reaction* – Bears who didn't move and rarely vocalized were put into this category, which accounted for 14 percent of our observations.

We saw significant differences in bear behavior between our first observation and subsequent observations. Bears who acted submissive or showed flight behavior during our initial observations frequently used threat behaviors when we approached them to administer the immobilizing drugs.

When we combined age classes we found no difference between the ways males and females behaved at trap sites. However, we did see differences between subadult and adult bears. Our data suggested that adult males were more aggressive than subadults. Rogers (4) and Barnes and Bray (27) found that large adult male black bears were dominant and frequently behaved aggressively at dumps. Stonorov and Stokes (28) and Egbert and Stokes (29) reported similar interactions among brown bears at fishing streams in Alaska.

Although these observations were made on snared bears, we believed they provided insights into how bears behave in the presence of threats. These data may help managers determine how vulnerable bears are to harvest and how bears might interact with people in areas where the two come in close contact.

LITERATURE CITED

1. POELKER, R. J., AND H. D. HARTWELL. 1973.
Black bear of Washington. State Game Dept. Biol. Bull. 14. 180pp.

2. AMSTRUP, S. C., AND J. J. BEECHAM. 1976.
Activity patterns of radio-collared black bears in Idaho. J. Wildl. Manage. 40: 340–48.

3. LINDZEY, F. G., AND E. C. MESLOW. 1977.
Home range and habitat use by black bears in southwestern Washington. J. Wildl. Manage. 41:413–25.

4. ROGERS, L. L. 1977.
Social relationships, movements, and population dynamics of black bears in northeastern Minnesota. Ph.D. thesis, University of Minnesota, Minneapolis. 194pp.

5. ALT, G. L., G. J. MATULA, JR., F. W. ALT, AND J. S. LINDZEY. 1980.
Dynamics of home range and movements of adult black bears in northeastern Pennsylvania. Int. Conf. Bear Res. and Manage. 5:131–36.

6. REYNOLDS, D. G., AND J. J. BEECHAM. 1980.
Home range activities and reproduction of black bears in west central Idaho. Int. Conf. Bear Res. and Manage. 4.

7. GARSHELIS, D. L., AND M. R. PELTON. 1980.
Activity of black bears in the Great Smoky Mountains National Park. J. Mamm. 61(1):8–19.

8. NOVICK, H. J., AND G. R. STEWART. 1982.
Home range and habitat preferences of black bears in the San Bernardino Mountains of southern California. California Fish and Game. 67:21–35.

9. YOUNG, B. F., AND R. L. RUFF. 1982.
Population dynamics and movements of black bears in east central Alberta. J. Wildl. Manage. 46(4):845–60.

10. SCHWARTZ, C. C., A. W. FRANZMANN, AND D. C. JOHNSON. 1983.
Black bear predation on moose. Final Rep. Alaska Dept. Fish and Game, Juneau. 135pp.

11. HERRNSTEIN, R. J. 1971.
Quantitative hedonism. J. Psychiatric Res. 8:399–412.

12. LINDZEY, F. G. 1976.
Black bear population ecology. PH.D. thesis, Oregon State University, Corvallis. 105pp.

13. PAYNE, N. F. 1975.
Unusual movements of Newfoundland black bears. J. Wildl. Manage. 39(4):812–13.

14. BEEMAN, L. E., AND M. R. PELTON. 1976.
Homing of black bears in the Great Smoky Mountains National Park. Int. Conf. Bear Res. and Manage. 3:87–95.

15. MCARTHUR, K. L. 1978.
Homing behavior of transplanted black bears, Glacier National Park. Natl. Park Serv. Prog. Rep., Glacier Natl. Park, West Glacier, Mont. 24pp.

16. ALT, G. L. 1980.
Rate of growth and size of Pennsylvania black bears. Pa. Game News 51(12):7–17.

17. MILLER, S. D., AND W. B. BALLARD. 1982.
Homing of transplanted Alaskan brown bears. J. Wildl. Manage. 46(4):869–76.

18. STICKLEY, A. R., JR. 1961.
A black bear tagging study in Virginia. Proc. Ann. Conf. S.E. Game and Fish Comm. 15:43–54.

19. ERICKSON, A. W., AND G. A. PETRIDES. 1964.
Population structure, movements, and mortality of tagged black bears in Michigan. Mich. State Univ. Res. Bull. 4:46–67.

20. JONKEL, C. J., AND I. M. COWAN. 1971.
The black bear in the spruce-fir forest. Wildl. Monogr. No. 27. 55pp.

21. BUNNELL, F. L., AND D. E. N. TAIT. 1981.
Population dynamics of bears – Implications. Pages 75–98. *In* C. W. Fowler and T. D. Smith, eds. Dynamics of large mammal populations. John Wiley and Sons, Ltd., New York, N.Y. 477pp.

22. EVELAND, J. F. 1973.
Population dynamics, movements, morphology, and habitat characteristics of black bears in Pennsylvania. Pennsylvania State University, University Park. 157pp.

23. HORN, H. S. 1968.
The adaptive significance of colonial nesting in the Brewer's blackbird (*Euphagus cyanocephalus*). Ecology. 49(4):682–94.

24. WEINS, J. A. 1976.
Population responses to patchy environments. Annu. Rev. Ecol. Syst. 7:81–120.

25. CRAIGHEAD, F. C., JR. 1971.
Biotelemetry research with grizzly bears and elk in Yellowstone National Park, Wyoming, 1965. Pages 49–62. *In* Natl. Geogr. Soc. Res. Rep. 1965 Proj.

26. MUNDY, K. R. D., AND D. R. FLOOK, 1973.
Background for managing grizzly bears in the national parks of Canada. Can. Wildl. Serv. Rep. Ser. 22. 35pp.

27. BARNES, V. G., AND O. E. BRAY. 1967.
Population characteristics and activities of black bears in Yellowstone National Park. Final Rep. Colo. Coop. Wildl. Res. Unit, Fort Collins. 196pp.

28. STONOROV, D. S., AND A. W. STOKES. 1972.
Social behavior of the Alaskan brown bear. Int. Conf. Bear Res. and Manage. 2:232–42.

29. EGBERT, A. L., AND A. W. STOKES. 1976.
The social behaviour of brown bears on an Alaskan salmon stream. *In* M. R. Pelton, J. W. Lentfer, and G. E. Folk, eds. Bears – Their biology and management. IUCN New Ser. 40:41–56.

TRYING TO DOCUMENT HOW BLACK BEARS USED THEIR
habitat involved long days in the field. We worked when the
bears were active, which included the hours before sunrise until just after sunset. And like the U.S. Postal Service, we
worked in rainy, sunny, hot, and cold weather because these
variables all caused differences in the ways bears used their
home ranges to meet their daily needs.

These day-to-day observations of radio-collared bears
gave us an opportunity to see some unique behaviors. Jeff will
probably never forget an early morning in June 1981, when he
was following the movements of an adult male, No. U-568.
As Jeff moved within 40 yards of the male, he realized he had
inadvertently gotten between No. U-568 and a radio-
collared female, No. U-810. Within minutes, a second radio-
collared male, No. U-570, showed up. Jeff soon realized that
he was in the middle of a threesome that included a female in
heat and two large males who were intent on making her ac-
quaintance. As Jeff tried discreetly to leave the area, he met a
fourth bear face-to-face in the brush. This male, No. U-510,
was an uncollared adult with a large, prominent scar across his
nose. We had captured him in a foot snare just a few days earlier. Jeff continued to back out of the area, but No. U-510 decided to follow him. When the bear was only 30 yards away,

Jeff began talking to him in hopes that his voice would frighten the bear away. It didn't. No. U-510 moved closer, within 10 yards, and Jeff began yelling at him, but to no avail. The bear stayed within 10 yards of Jeff until Jeff reached the road and the safety of his truck. Slightly unnerved by the experience, Jeff decided to call it a day, returned to camp, and seriously considered a fresh haircut and shave.

JOHN BEECHAM

Our research showed that black bears used their habitat in predictable ways. From the time they left their dens each spring until they entered them the following fall, bear movements were dictated by the distribution and availability of food, but security needs were also important. Bears chose secure areas when they bedded down during the day or denned for the winter.

The major goal of our habitat research was to document black bear habitat use patterns at Council and Priest Lake, and use this information to develop timber management guidelines that would help maintain or enhance bear habitat. Our specific objectives were to (1) determine physical and environmental factors that affected habitat use by black bears, (2) document the bears' food habits, (3) identify relationships between habitat use and black bear food plants (their variety, abundance, and developmental stages), (4) identify the effects of different silvicultural practices on black bear habitat use, and (5) prepare timber management guidelines.

Black bear populations occur throughout much of Idaho, but are largely confined to coniferous forests in the northern two-thirds of the state and isolated areas of eastern Idaho (1). There are many competing uses for these forested areas including mining, mineral and oil exploration, recreation, water development, livestock grazing, and timber production. Of these, timber production and the associated increase in road access to forested areas probably have the greatest effect on black bears. With the increased popularity of black bears as game animals and increasing demands on forestlands for timber and other uses, we need research on habitat use to ensure the welfare of black bears and their habitat.

Habitat use studies have been conducted in Montana (2), Alberta (3), Arizona (16), California (4, 5), and Virginia (6). Other habitat use studies related directly to timber management and bears have been done in Montana (7), Idaho (8, 9), and Washington (10). Because black bear populations are unique products of specific habitat factors that influence population dynamics, social organization, reproductive potential, food habits, and availability of suitable den sites (11), data from different geographic areas may not be applicable to Idaho.

COUNCIL STUDY AREA

On the Council study area, we examined the bears' overall and seasonal habitat use patterns and compared habitat use by bears with and without cubs, and by type of bear activity (Plate 5). We also noted whether bears preferred certain aspects, topographic classes, elevations, slopes and distance to roads, water, and cover.

Ten adult female bears were radio-collared to examine their habitat use patterns. We located these bears 640 times in 1982 and 1983, obtaining 197 visual observations (31 percent), 379 close observations (59 percent), 53 close triangulations (8 percent), and 11 triangulations (2 percent).

Over 90 percent of the time, radio-collared bears used three of the eight major habitat types or series occurring on the Council study area: (1) grand fir/huckleberry, (2) grand fir/mountain maple, and (3) Douglas fir/ninebark (12). Because habitat types describe climax vegetation rather than existing vegetative cover, we used cover types instead of habitat types to describe habitat use patterns (Table 7–1).

OVERALL HABITAT USE

With all seasons and activities combined, bears most often chose timber, open timber/shrubfield, and riparian habitats, and less often selected meadow, road, rock/scree, sagebrush/grass, and clear-cut habitats (see Table 7–2 for a breakdown of habitat use by season and activity).

Timber was the most frequently used habitat on the Council

Table 7–1. Cover types used on the Priest Lake and Council study areas

Timber	Unlogged stand of timber with canopy closure >60%
Open Timber	Unlogged stand of timber with canopy closure >30% but <60%. Undergrowth dominated by grasses and forbs
Open Timber/ Shrubfield	Unlogged stand of timber with canopy closure >30% but <60%. Undergrowth dominated by shrubs
Riparian	Streamside or moist areas with well-developed mesic vegetation
Aspen	Stands with dense overstory dominated by quaking aspen
Shrubfield	Unlogged areas with timber canopy closure <30%. Undergrowth dominated by shrubs
Meadow	Open sites dominated by grasses and forbs
Rock/Talus	Extensive areas of exposed bedrock or rock slides
Sagebrush/ Grass	Open areas dominated by big sagebrush, grasses, and forbs
Roads	Cleared or graded areas that are not blocked to vehicular travel
Clear-cut	Logged areas with overstory completely removed. Dominated by shrubs
Selection Cut/ Shrubfield	Logged areas with overstory <30% and undergrowth dominated by shrubs
Selection Cut/ Open Timber	Logged areas with overstory >30% but <60%. Undergrowth dominated by shrubs
Selection Cut/ Timbered	Logged areas with overstory >60% and sparse undergrowth dominated by shrubs and forbs
Slabrock	Naturally open to sparsely timbered sites with exposed blocks of glaciated bedrock
Snowchute/ Shrubfield	Naturally open sites on steep slopes at high elevations created by periodic movements of snow

Table 7–2. Availability and use (%) of cover types by season and activity categories for 10 female black bears near Council, Idaho, 1982–83

Cover Type	Random Availability $(n = 489)^b$	Spring Use $(n = 151)$	Summer/ Fall Use $(n = 483)$	Feeding Use $(n = 123)$	Bedding Use $(n = 281)$
Timber	13.9	43.6 + [a]	37.9 +	12.3	56.9 +
Open Timber	10.6	12.8	5.0–	5.7	5.0–
Open Timber/ Shrubfield	7.4	8.3	13.9 +	15.6	11.4
Riparian	0.4	1.3	2.5 +	1.6	0.4
Aspen	0.8	1.3	3.3 +	3.3	3.2
Shrubfield	4.7	0.6–	7.9	14.8 +	5.0
Meadow	6.7	3.2	0.0–	3.3	0.0–
Rock/Talus	1.6	0.0–	0.0–	0.0–	0.0–
Sagebrush/Grass	17.8	3.2–	0.6–	2.5–	0.0–
Roads	3.5	1.3	0.4–	0.8	0.0–
Clearcut	2.7	0.6	0.2–	0.8	0.4–
Selection Cut/ Shrubfield	5.7	7.1	8.9	16.4 +	3.6
Selection Cut/ Open Timber	20.2	15.4	16.6	21.3	10.3–
Selection Cut/ Timber	3.9	1.3	2.9	1.6	3.9

[a] + indicates use greater than availability and – indicates use less than availability (P <0.10).
[b] n = number of radio locations.

study area (39 percent). We classified 65 percent of our radio locations as bedding sites, which were typically located on steep, north- or east-facing slopes in dense timber with little ground cover. At Priest Lake, the bears' preferred habitats[1] for bedding sites were timbered areas with a sparse understory, even when the bears spent most of their time in shrub-dominated selection cuts (8). In California, Kellyhouse (4) reported that bears used mixed coniferous forests with an estimated overhead canopy coverage of 68 percent for traveling, resting, and escape during all seasons.

At Council, timber habitats were used in proportion to their availability[2] by feeding bears. Important bear foods like huckleberry, buffaloberry, twinberry, and serviceberry grew in timber habitats but in lower amounts than in the more open habitats (9). Seasonally important forbs and grasses were also found in timbered areas.

The open timber/shrubfield habitat was the second most popular habitat. Nearly 70 percent of the bears' use of open timbered areas occurred in the spring, when bears fed predominantly on grasses and forbs. By midsummer these sites were dry, and food plants weren't as palatable to bears (13, 14). Bears without cubs used the open timber habitat in spring, but bears with newborn cubs remained in dense timber stands and seldom ventured into open areas. We believe that females with cubs selected dense timber stands because they provided greater cover.

When open timber habitats were combined with shrubfields, however, bears with cubs used them heavily. In summer and fall this habitat offered bears a variety of fall food items, and its dense cover provided bedding and security cover for females with cubs.

For all bears, shrubfields were very important sources of berries in summer and fall; over 90 percent of the bear use in this habitat occurred after August. Hawthorn, bitter cherry, and chokecherry were the most sought-after berry species in shrubfields.

We expected riparian areas to be more popular with bears, but they used riparian areas only in proportion to their availability. Perhaps this was because riparian areas were scarce or weather conditions were wet during the study. Riparian habitats had well-developed *mesic vegetation*, and we believe that during dry years, these areas may be more important to bears. Black bears selected riparian areas as feeding sites in northern Idaho (15) and as feeding areas and traveling corridors in California (4) and Arizona (16).

Black bear use of meadows on the Council study area was heaviest during the spring, but this use was lower than we anticipated. We believe our findings understated the overall importance of meadows to west-central Idaho bears during the spring because female bears with cubs didn't select meadows. In 1982, eight of the radio-collared bears were without cubs, but few of these bears were captured in time for us to observe their spring habitat use. However, we often saw noncollared bears feeding in meadows, and we observed radio-collared bears without cubs using meadows when we returned to the study area in the spring of 1984. On the study area, meadow habitats provided a wide range of forbs and grasses important to the bears' spring diet (17). Meadows were also very important spring feeding sites for California black bears (4).

During all seasons and all activities, bears rarely selected rock/talus and sagebrush/grass habitats. Although the habitats had food plants used by some bears in the spring, other habitats apparently provided a greater abundance of preferred foods and greater security cover.

Logged areas made up over 30 percent of the available habitats on the study area and included 28 percent of our bear radio locations. Use of these areas depended on whether the areas were selectively cut or clear-cut. Selection cuts provide a wide variety of bear foods: This logging practice opens up the forest canopy, allowing more bear foods to grow, but doesn't *scarify* the soil surface, which can slow the growth of food plants. In Arizona, Mollohan (16) found that adult female bears preferred 50-to-60-year-old selection cuts with dense shrub and grass understories.

Security cover is clearly important to bears. We believed that bears chose open timber and selection-cut habitats because of the greater security provided by increased overhead canopy cover. Zager (18) emphasized the importance of leaving timbered areas next to roads as a buffer and leaving small amounts of residual cover in clear-cuts. He also noted the importance of timber "stringers," or buffer zones, along travel routes. Young and Ruff (19) reported that bears used heavy spruce cover more often during the fall hunting season, presumably because it offered better security cover.

Clear-cut habitats made up a small portion of the study area (3 percent), and we saw bears in them only twice during the study. The clear-cuts were less than 8 years old, and although some bear foods grew on these sites, the foods most commonly found in scats didn't appear to be as abundant in the clear-cuts as in more mature timber stands. Bears in western Washington selected clear-cuts 18 to 25 years old, but stayed out of clear-cuts 9 to 14 years old (10). In northern Montana, Jonkel and Cowan (2) found that black bears seldom used recently logged areas, but used a 10-year-old clear-cut as much as surrounding areas.

SEASONAL HABITAT USE

Habitat use varied significantly between seasons. In addition, spring (April-June) and summer/fall (July-November) habitat use differed significantly from availability. During both seasons, timber and selection cuts were the bears' preferred habitats, while clear-cut areas were avoided habitats.[3] Timber accounted for 249 (28.9 percent) of the spring radio locations, 310 (21.6 percent) of the summer locations, and 117 (36.8 percent) of the fall locations. In the spring, bears avoided shrubfield, rock/talus, and sagebrush habitats, but used other habitats in proportion to their availability. In summer/fall, they avoided open timber, meadow, rock/talus, sagebrush/grass, and road habitats, and preferred open timber/ shrubfield, riparian, and aspen habitats. They used the remaining habitats in proportion to their availability.

HABITAT USE BY BEARS WITH AND WITHOUT CUBS

Habitat use varied significantly between females with cubs and females without cubs. Bears with cubs preferred open timber/ shrubfield and aspen habitats, and avoided open timber and road habitats. Neither group, however, used all habitats in proportion to their availability. Both groups selected timber habitats and shunned meadow, rock/talus, clear-cut, and sagebrush/grass habitats.

HABITAT USE BY TYPE OF ACTIVITY

We classified black bear activities as feeding (20 percent), bedding (44 percent), traveling (8 percent), denning (3 percent), and unknown (26 percent).

Bear use of bedding habitats differed significantly from availability. For bedding, bears frequently used timbered habitats and avoided open timber, meadow, rock/talus, sagebrush/grass, road, clear-cut, and selection-cut/open timber habitats.

During the spring, bears used open timber habitats for feeding on newly emerged grasses and forbs. Photo by Jeff Rohlman

Use of feeding habitats also varied significantly from availability. Bears preferred selection-cut/shrubfield and shrubfield habitats, and stayed out of clear-cut, rock/talus, and sagebrush/grass habitats.

Because of small sample sizes for traveling and denning locations, we couldn't analyze those data.

USE OF ASPECT

Bears didn't use the various aspects in proportion to their availability except when feeding (Table 7–3). Aspect use didn't differ significantly between seasons or for bears with and without cubs, but did vary between feeding and bedding locations. When feeding, female black bears used aspects in proportion to availability. At bedding time, bears selected north aspects and avoided west and south aspects.

Bears also used north-facing slopes as travel corridors. On the Council study area, finger ridges extend in a westerly direction and

Table 7–3. Use of aspect, topography, roads, and water (%) by season and activity categories for 10 female black bears near Council, Idaho, 1982–83

Variable	Random Availability	Spring Use	Summer/Fall Use	Feeding Use	Bedding Use
Aspect					
North	27.6	53.5 +[a]	51.4 +	37.0	59.0 +
East	15.1	12.9	15.4	16.5	14.5
South	20.9	5.8–	7.9–	16.5	5.7–
West	36.4	27.7	25.3–	29.9	20.8–
Topography					
Ridgetop	9.6	5.9	3.1–	4.1–	3.2–
Upper slope	21.9	15.8	15.5–	16.3	16.5
Midslope	47.5	43.4	39.6–	29.3–	44.4
Lower slope	13.5	25.0 +	27.0 +	24.4	28.3 +
Bench-flat	5.1	6.6	7.3	17.1 +	3.9
Stream bottom	2.3	3.3	7.3 +	8.9 +	3.6
Distance to road					
0–164 feet	18.8	6.1–	10.0–	13.7	4.3–
>164 feet	81.2	94.0 +	89.8 +	85.4	95.7 +
Distance to water					
0–328 feet	34.0	59.2 +	60.5 +	64.2 +	57.9 +
>328 feet	65.8	40.7–	39.5–	35.8–	42.0–

[a] + indicates use greater than availability and – indicates use less than availability (P <0.10)

at lower elevations were only timbered on the north aspects. Hawthorn shrubfields were usually found on lower slopes and southern exposures. In late summer and fall, bears would travel down the north side of these finger ridges and slip into shrubfields on the opposite, south-facing slopes to feed.

USE OF TOPOGRAPHY

Bears didn't use topographic classes randomly (Table 7–3). Overall, bears avoided ridgetops and upper slopes, and preferred lower slopes. These lower hillsides were wetter than other areas, and provided cover as well as food.

In the spring, bears used all topographic classes in proportion to availability except lower slopes, which they preferred. During the summer/fall period, the bears chose lower slopes and stream bottoms, and didn't select ridgetops and upper and midslopes.

Feeding bears avoided ridgetops and steeper midslopes, preferring benches, flats, and stream bottoms. These sites were typically wetter than other areas, and most shrubfields were located on this type of topography. Shrub species that provided bear foods were present on steeper midslopes, but small benches and flats on midslopes supported the densest shrub stands. In the spring, upper slopes and ridgetops were used as feeding areas, but they were the first landforms to dry out, so they provided food for a limited time.

For bedding sites, bears preferred lower slopes and avoided ridgetops.

Female black bears with cubs avoided ridgetops and used upper and midslopes, benches, and flats in proportion to their availability. Bears without cubs selected lower slopes, spending little time on ridgetops and upper slopes.

HABITAT USE AND DISTANCE TO ROADS

All roads in the study area were one- and two-lane, gravel- or dirt-surfaced forest roads. Overall, bears preferred to stay more than 50 yards from roads (Table 7–3). This pattern was consistent for bears with or without cubs during the spring and summer/fall seasons.

In the spring, however, feeding bears and bears without cubs did use roads in proportion to their availability. When distance to roads was used as the criterion, only feeding bears used areas within 50 yards of roads in proportion to their availability. We weren't sure why feeding bears were willing to risk human contact, but we did know that females in the Council bear population hadn't been hunted since 1976. Although bears would readily run or climb trees at the sight or smell of humans, it's possible some Council bears were becoming accustomed to humans. Bedded bears and bears with cubs avoided areas less than 50 yards from roads, probably because these areas lacked security cover.

Roads on the Council study area weren't normally used as

travel routes because bears could easily move through forested areas that generally lacked thick undergrowth. Manville (20) reported that Michigan black bears used oil pipeline right-of-ways, oil well service lanes, and lumber roads as travel routes. Increased human access along these roads made bears more vulnerable to hunters.

HABITAT USE AND DISTANCE TO WATER

Overall, in the spring and summer/fall periods, bears chose areas within 100 yards of water (Table 7–3). During this study, we didn't see bears drinking water, but the hot, dry summers could make water important. The bears' close association with water could also be related to the greater growth of bear foods in wetter sites. Mollohan (16) found that bears in Arizona were within one-quarter mile of water at over 60 percent of the bedding sites and 50 percent of the feeding sites.

HABITAT USE AND DISTANCE TO COVER

Bears showed a strong preference for areas with horizontal cover. Female bears especially preferred to be in cover and avoided all areas away from cover except when eating. Feeding bears used areas less than 25 yards from cover in proportion to availability.

The importance of cover to bears has been stated by other authors (21, 22, 23, 15). Bears were willing to leave cover if food was available and they couldn't detect human activity. We frequently saw bears in open meadows during the spring and in shrubfields in fall; however, we were usually a long distance from them.

USE OF ELEVATION AND SLOPE

The elevations of bear radio locations varied significantly by month, depending on the availability and ripeness of foods. The general feeding pattern for bears in North America is spring use of forbs and grasses, and summer/fall use of hard or soft *mast* (24–28).

This general pattern occurred in Idaho, and elevational movement by Council bears during this study was associated with the quality and quantity of important foods, as reported by Amstrup and Beecham (13) and Reynolds and Beecham (14). Bears fed on low-elevation grasses and forbs in spring; as grasses and forbs matured at higher elevations, bears moved to higher ground. During July, before huckleberries and buffaloberries were ripe, bears fed extensively on ants in selection cuts at middle elevations.

Bears followed one of two elevational use patterns in late summer/early fall. Some bears fed on huckleberries and buffaloberries until they were no longer available, and then went to lower elevations to eat chokecherries, bitter cherries, and hawthorn berries. Bears whose home ranges didn't include large stands of huckleberries or buffaloberries or who remained at lower elevations fed on chokecherries, bitter cherries, and hawthorn berries. We didn't see these bears moving up or down in search of huckleberries.

The slopes used by bears varied with their activities. Bears used steep, northern aspects as bedding sites and used gentler areas for feeding. We believe that bears selected these sites not on the basis of gradient, but because of vegetative characteristics, which were affected by slope and other topographic factors.

Bears who were traveling and denning used all slopes in proportion to availability.

PRIEST LAKE STUDY AREA

At Priest Lake, we looked at overall and seasonal habitat use patterns and determined how habitat use differed between males and females. We radio-collared eight bears and located them 684 times.

OVERALL HABITAT USE

On our studies at Priest Lake, bears preferred selection cuts and timber habitats, avoided clear-cuts, and didn't use riparian zones, meadows, and avalanche chutes as much as we expected. We found that selection cuts were the most important habitats used by bears at Priest Lake, and attributed this to the habitat diversity, high concentration of key bear foods, and cover found in those habitats.

Habitat diversity appeared to be an important element of black bear habitats (2, 29, 30). In the past, habitat diversity was associated with wildfire. Zager (18) stated that "wildfire historically burns in a mosaic pattern, which results in an interspersion of habitats, and ensures heterogeneity [a mixture] of environments." Wildfire and logging practices that simulated wildfire, such as selection cutting, were instrumental in creating and maintaining habitat diversity for bears at Priest Lake.

Wildfire and selection cuts also increased the productivity and abundance of bear foods. Black bears in the Pacific Northwest use many berry-producing shrubs commonly found in seral plant communities (2, 31, 10). Wildfire played a major role in creating and maintaining these seral communities by removing the tree canopy. As more sunlight reached the forest floor, the growth and productivity of shrubs increased. Martinka (32), Minore (33), and Martin (34) documented greater and more consistent huckleberry production on burned sites than under a mature forest canopy. Serviceberry growth also increased after a wildfire (35, 36). Hemmer (37) found serviceberry most dense in old burns where adequate moisture was available. In addition, Franklin and Dyrness (36) characterized mountain ash as a pioneer species that resprouts vigorously after a fire.

Selection cuts may also help create and maintain the seral communities that provide berry crops. Studies by Hagar (38) and Ahlgren (39) showed that wildlife responses to logging were similar to responses seen after wildfire, suggesting that logging and wildfire may have similar effects on wildlife. Zager (18) stated that certain timber harvest methods may influence grizzly bear habitat in the same manner as wildfire. He added that the open canopy created by wildfire may be simulated by logging. In several studies, higher concentrations of bear foods were commonly found in seral plant communities after logging (40, 10, 18). Mealey (41) reported that selection cuts were the most important bear habitats (based on the availability of key bear foods) of the logged habitats in his northwestern Montana study area.

Assuming that wildfire and selective logging have similar ecological effects on bear habitat, the productivity of key bear foods should be high on selection cuts at Priest Lake. Commercial logging began there in the early 1900s, and has increased since the 1940s. From 1940 to 1960, most timber was selectively logged (42).

Thus, most selection cuts on the study area were 20 to 40 years old. Zager (18) noted that in northwestern Montana, the canopy cover of shrubs considered to be key grizzly bear food was higher on sites burned by wildfire 35 to 70 years earlier. These same shrubs were also important to black bears at Priest Lake. Martin (34) found that the productivity of huckleberries on high-elevation sites was highest in stands that were burned 25 to 60 years before.

The concentration of highly productive, fruit-bearing shrubs partly explains why black bears preferred selection cuts during the summer and fall, when their diet consisted mainly of berries. However, black bears at Priest Lake also chose selection cuts during the spring, when their diet consisted primarily of grasses. On several occasions, we saw radio-collared bears feeding on pinegrass, often in areas with no other spring foods. Pinegrass reproduces vegetatively[4], and often increases when the tree canopy is removed (43). We surmised that the rapid growth of pinegrass after selection logging was a major factor resulting in high bear use of this habitat during the spring.

In addition, the scarification of soils after logging may have influenced bear use of selection cuts during the spring. Jonkel and Cowan (2) and Zager (18) found that certain species of grasses normally uncommon in old-growth stands and old burns were typically found in disturbed soil along roads. They also noted that horsetails, common in the spring diet of Priest Lake black bears, grew more vigorously after logging and were found along moist skid roads.

Black bears may also prefer selection cuts in the spring because those habitats contain huckleberry flowers, part of the bears' spring diet. We found huckleberry leaves and flowers in 45 percent and 29 percent of the scats collected in May and June 1981 (44). Several of these scats were composed entirely of huckleberry leaves. It's hard to find delicate plant tissues such as flower petals in bear scats because these tissues are easily digested. Based on the occurrence of huckleberry leaves, stems, and some flower parts in scats, we suspected that bears at Priest Lake were feeding consistently on huckleberry flowers during spring, but the flower parts were digested and didn't show up in our food habits analysis. In June 1981 we watched a radio-collared male black bear as he fed on huckleberry flowers. The abundance of huckleberries in selection cuts provided bears with an almost unlimited source of huckleberry flowers during the spring and resulted in increased bear use of this habitat.

Selection cuts provided not only food but also cover. Several studies showed the importance of cover in black bear habitat selection (10, 5, 9). Lindzey and Meslow (22, 23) found that bears selected older cutting units with more cover even though younger cutting units provided more food. The shrub layer in selection cuts at Priest Lake was extremely dense and often seemed impenetrable. On many occasions we could hear, but not see, radio-collared bears in older selection cuts because they were hidden by the dense understory.

Selection cuts at Priest Lake also contained an average of 220 trees/acre scattered singly or in small groups. Herrero (21) considered trees significant in the daily existence of black bears, particularly because they allowed the bears to climb to safety. Lindzey and Meslow (10) reported that a significant percentage of the bears they located in clear-cuts were along the edges, suggesting that trees were important to bears for safety.

We concurred with Lindzey and Meslow (22, 23) that hiding and security cover and food availability were important to habitat selection. Black bears used selection cuts at Priest Lake extensively because they provided the best combination of food and cover of all available habitats.

Timber was the second-most used habitat at Priest Lake, although its use was in proportion to availability. Barnes and Bray (45), citing earlier studies, also noted that forested areas were an important component of black bear habitat. Herrero (21) discussed the importance of timber habitats in black bear evolution. Lindzey and Meslow (10) noted that over half of the black bears they spotted in timber were on forest edges, suggesting that timber habitats provided resting and escape cover for bears who normally foraged in clear-cuts. We suspected that timbered areas function primarily as cover for black bears at Priest Lake. The fact that bears used timber habitats more during the spring and fall, seasons in which they spent most of the day bedded, supported this conclusion.

We expected greater spring use of riparian zones, wet meadows, and avalanche chutes by black bears than we observed; bears used these habitats in proportion to their availability. Because riparian zones were normally narrow and shaded by the forest canopy, their understories contained plant species similar to those of nearby habitats (46). Therefore, riparian zones at Priest Lake offered bears little more in the way of succulent spring vegetation than did adjacent areas, so were used less than we expected.

Black bear use of wet meadows and avalanche chutes was also less than we anticipated. We believed bears would make substantial use of these habitats because of the succulent foods available (46, 18). However, wet meadows and avalanche chutes covered only about 0.2 percent of the study area and were normally located at high elevations in inaccessible areas. As a result, we may have overestimated their use by black bears.

Clear-cut areas weren't popular with radio-collared bears, and they avoided them during the spring, summer, and fall. Clear-cutting has been the dominant timber harvesting method employed at Priest Lake since the 1960s. In most cases, loggers burned timber slash after harvest and scarified some sites extensively to enhance the growth of new trees (47). Scarification of soils damages the rhizomes and root crowns of vegetatively reproducing shrubs such as huckleberry (18). Clear-cuts were generally moderately to heavily scarified at Priest Lake (47) and had an average huckleberry cover class value of 1 (1 to 5 percent of the area was covered with huckleberries) compared with an average cover class value of 2 (5 to 25 percent) for selection cuts. Zager (47) estimated it would take 10 to 20 years for minimally scarified sites in the Priest Lake area to recover sufficiently to provide food and cover for bears; extensively scarified sites would require much more recovery time. The oldest clear-cuts on the study area were about 20 years old, but most clear-cuts were more recent. Because these clear-cuts weren't old enough to provide adequate cover and food, we saw little use of them by bears.

Meadow habitats weren't as popular with bears as we had expected. Photo by Jeff Rohlman

Jonkel and Cowan (2) reported that Montana black bears didn't use young-aged clear-cuts or areas that were recently logged. Lindzey and Meslow (10) found that black bears used 6-to-11-year-old clear-cuts in Washington less than expected, but used 15-to-24-year-old clear-cuts more than expected. Martin (34) sampled eight different habitats in northwestern Montana and found that cover and huckleberry volumes were lowest in scarified clear-cuts. Tisch (46) reported that recent clear-cuts provided very little black bear food in the Whitefish Range of northwestern Montana.

Grasses and forbs, which were important to bears during the spring, usually became more plentiful after scarification (18). Although grasses and forbs were abundant in scarified clear-cuts at Priest Lake, black bears didn't use these sites in spring, possibly because there was little cover in these units or because the plant species that increased after logging weren't those preferred by bears. This suggested that food availability and cover were important requirements for black bear habitat selection.

SEASONAL HABITAT USE

Black bears at Priest Lake showed seasonal changes in habitat use patterns. We believe that seasonal variation in food availability, resulting from key foods maturing at different times, was responsible for these changes (15). Jonkel and Cowan (2), Kellyhouse (4), and Novick and Stewart (5) observed similar habitat use patterns in other areas.

HABITAT USE BY MALES AND FEMALES

Black bear habitat use differed significantly by sex. The most dramatic difference was that females used timber habitats more than males. Herrero (21), discussing the way cub production had evolved in black bears, stated that "females with cubs are reluctant to leave [the safety of] trees." Perhaps this behavioral trait caused the difference in use of timber habitats by male and female bears at Priest Lake. It may not be the only factor, however. Although three of four radio-collared females were accompanied by cubs in 1981, their habitat use wasn't significantly different from that seen in 1980, when those females lacked cubs.

Male black bears tended to use roads in proportion to availability, while female black bears avoided them. This tendency has strong management implications because it means that male black bears are more susceptible to harvest. Female bears may have shunned roads due to a maternal instinct to avoid less secure areas. Zager (18) documented a female grizzly bear with cubs avoiding roads on his study area in northwestern Montana. If roads serve as travel routes, male black bears may use roads more often than females because males are more mobile (13, 14).

MANAGEMENT RECOMMENDATIONS

Timber harvest methods and postlogging site treatments have changed radically in the past 20 years, shifting from high-grade, selection-cutting techniques to clear-cutting. A large portion of the important bear habitat on the Council and Priest Lake study areas was a result of selective logging activities in the past. However, more recent forest management plans called for increased use of clear-cutting. Intensive postlogging site treatments are planned that include bulldozing slash into rows and burning it, along with extensive soil scarification. In the long run, this shift in timber management will harm black bear habitat and, ultimately, bear populations.

Timber managers can avoid many of these detrimental impacts by modifying their clear-cutting procedures. Since clear-cutting does shift vegetation to early and, eventually, midseral stages that produce large quantities of bear foods, clear-cuts should be *broadcast-burned* or left untreated after logging (10, 7, 48, 49) rather than *windrow-* or *jackpot-burned* and bulldozer-scarified.

To maintain bear populations during the 10 to 35 years (49, 7, and 9) that clear-cuts will be marginally suitable for bears, clearcuts should be small, irregularly shaped, and cut on a harvest schedule that will maintain an acceptable mix of different-aged cutting units (18, 15). The mixture of these units will influence the density and distribution of bears in the area. Specific sites within each cutting unit, including dense timber stands on north aspects and strips along streams and roads, should be left alone to enhance their use by females with cubs after shrubs have recovered.

Land managers should also develop a cooperative road access management plan for timber harvest on private and public lands. This plan should incorporate agreements on road placement (use the minimum needed to accomplish cutting objectives), road standards (use minimal-quality roads that can withstand necessary traffic, but maintain bear screening cover, etc.), and road access (allow minimal traffic consistent with logging needs and public access requirements).

Additionally, timber managers should minimize disturbances to black bears by coordinating logging activities with the seasonal habitat use patterns of bears. For instance, logging activities near high-elevation shrubfields should be avoided in late summer. To maintain or enhance black bear food production, managers should use a *let-burn fire management policy* or *prescribed burns* where feasible.

In many situations, responsible logging practices help black bears because they enable habitats to support more bears. However, we must recognize that increased human access into black bear habitats makes bears more vulnerable to hunters. This factor may offset the positive benefits of habitat improvement. Convincing sportsmen, U.S. Forest Service personnel, timber companies, and private landholders to develop and use comprehensive logging and access management plans for black bear habitats will be a challenging task. Competing public and private interests must accept some rational compromise to accommodate the biological needs of wildlife and the social and economic needs of the state.

Specific timber management recommendations for west-central and northern Idaho follow.

COUNCIL

We made the following timber management recommendations to maintain or enhance bear habitat, based on our observations of female black bear habitat use in west-central Idaho. This area is much drier than northern Idaho and removal of the timber overstory caused significant changes in plant diversity and the availability of bear foods. Specific management recommendations include:

1. Minimize soil disturbance, especially in areas where berry-producing shrubs are abundant.
2. Harvest timber on north aspects and along streams using selection cuts to maintain bear security cover.
3. Maintain dense pole-sized timber stands on north- and east-facing slopes as bedding areas.
4. Retain mature trees in logged areas to enhance their use by female bears with cubs after berry-producing shrubs have recovered.
5. Protect aspen stands.
6. Design clear-cuts with irregular borders to maximize security cover; maintain patches or strips of timber as travel routes; and scatter or broadcast-burn slash or leave it untreated.
7. Minimize the impact of timber harvest on a given drainage by designing timber harvest plans to provide a mixture of different-aged cuts next to one another.

In addition to these recommendations, we suggested that land managers try to maintain movement corridors between intensively managed areas to assist dispersing subadult black bears.

PRIEST LAKE

We made the following timber management recommendations to maintain or enhance bear habitat in northern Idaho. This list is based on our data from Priest Lake and is not all-inclusive. Our recommendations include:

1. Maintain or improve black bear food production on logged sites. Broadcast-burn slash or leave it untreated, and minimize soil scarification to prevent damage to vegetatively reproducing food plants.
2. Retain timber near potential or existing feeding areas. Timber is important to bears as resting, escape, and security cover. To retain cover, create *leave patches* and *leave strips* within cutting units. Clear-cuts should have irregular borders because they provide greater cover.
3. Coordinate logging activities to maintain an acceptable mix of different-aged cutting units. This mixture will influence the density and dispersion of bears in the area.
4. Leave mature, standing trees in cutting units so bears can escape danger by climbing.

LITERATURE CITED

1. BEECHAM, J. 1983.
Population characteristics of black bears in west central Idaho. J. Wildl. Manage. 47(2):402–12.
2. JONKEL, C. J., AND I. M. COWAN. 1971.
The black bear in the spruce-fir forest. Wildl. Monogr. No. 27. 55pp.
3. FULLER, T. K., AND L. B. KEITH. 1980.
Summer ranges, cover type use, and denning of black bears near Fort McMurray, Alberta, Canada. Field-Nat. 94:80–3.
4. KELLYHOUSE, D. 1980.
Habitat utilization by black bears in northern California. Int. Conf. Bear Res. Manage. 4:221–27.

5. NOVICK, H. J., AND G. R. STEWART. 1982.
Home range and habitat preferences of black bears in the San Bernardino Mountains of southern California. California Fish and Game. 67:21–35.

6. VAUGHAN, M. R., E. J. JONES, D. W. CARNEY, AND N. GARNER. 1983.
Seasonal habitat use and home range of black bears in Shenandoah National Park: First Ann. Prog. Rep., 1 July 1983.

7. ZAGER, P. E., C. JONKEL, AND J. HABECK. 1983.
Logging and wildfire influence on grizzly bear habitat in northwestern Montana. Int. Conf. Bear Res. Manage. 5:124–32.

8. YOUNG, D. D., AND J. J. BEECHAM. 1986.
Black bear habitat use at Priest Lake, Idaho. Int. Conf. Bear Res. Manage. 6:73–80.

9. UNSWORTH, J. W., J. J. BEECHAM, AND L. R. IRBY. 1989.
Female black bear habitat use in west central Idaho. J. Wildl. Manage. 53(3):668–73.

10. LINDZEY, F. G., AND E. C. MESLOW. 1977.
Home range and habitat use by black bears in southwestern Washington. J. Wildl. Manage. 41:413–25.

11. BEECHAM, J. 1980.
Population characteristics, denning, and growth patterns of black bears in Idaho. Ph.D. thesis, University of Montana, Missoula. 101pp.

12. DAUBENMIRE, R., AND J. DAUBENMIRE. 1968.
Forest vegetation of eastern Washington and northern Idaho. Wash. Agric. Exp. Stn. Tech. Bull. 60. 104pp.

13. AMSTRUP, S. C., AND J. J. BEECHAM. 1976.
Activity patterns of radio-collared black bears in Idaho. J. Wildl. Manage. 40:340–48.

14. REYNOLDS, D. G., AND J. J. BEECHAM. 1980.
Home range activities and reproduction of black bears in west central Idaho. Int. Conf. Bear Res. and Manage. 4.

15. YOUNG, D. D. 1984.
Black bear habitat use at Priest Lake, Idaho. M.S. thesis, University of Montana, Missoula. 66pp.

16. MOLLOHAN, C. 1982.
Black bear habitat research. Final Rep. No. 0182: 1 September 1982. Arizona Fish and Game, Phoenix.

17. BEECHAM, J. 1976.
Black bear ecology. Job Prog. Rep. Idaho Dept. Fish and Game, Boise. 34pp.

18. ZAGER, P. E. 1980.
The influence of logging and wildfire on grizzly bear habitat in northwestern Montana. Ph.D. dissertation, University of Montana, Missoula. 131pp.

19. YOUNG, B. F., AND R. L. RUFF. 1982.
Population dynamics and movements of black bears in east central Alberta. J. Wildl. Manage. 46(4):845–60.

20. MANVILLE, A. M. 1983.
Human impact on the black bear population in Michigan's Lower Peninsula. Int. Conf. Bear Res. and Manage. 5:20–33.

21. HERRERO, S. 1972.
Aspects of evolution and adaptation in American black bears and brown and grizzly bears of North America. Int. Conf. Bear Res. and Manage. 2:221–31.

22. LINDZEY, F. G., AND E. C. MESLOW. 1976A.
Winter dormancy in black bears in southwestern Washington. J. Wildl. Manage. 40:408–15.

23. LINDZEY, F. G., AND E. C. MESLOW. 1976B.
Characteristics of black bear dens on Long Island, Washington. Northwest Sci. 50:236–42.

24. BEEMAN, L. E., AND M. R. PELTON. 1980.
Seasonal foods and the feeding ecology of black bears in the Smoky Mountains. Int. Conf. Bear Res. and Manage. 4:141–48.

25. BENNET, L. J., P. F. ENGLISH, AND R. L. WATTS. 1943.
The food habits of the black bear in Pennsylvania. J. Mammal. 24:24–31.

26. GRABER, D. M., AND M. WHITE. 1983.
Black bear food habits in Yosemite National Park. Int. Conf. Bear Res. and Manage. 5:1–10.

27. GRENFELL, W. E., AND A. J. BRODY. 1983.
Seasonal foods of the black bears in Tahoe National Forest, California. California Fish and Game. 69:132–50.

28. LANDERS, J. L., R. J. HAMILTON, A. S. JOHNSON, AND R. L. MARCHINTON. 1979.
Foods and habitat of black bears in southeastern North Carolina. J. Wildl. Manage. 43:143–53.

29. KEMP, G. A. 1979.
Proceedings of the workshop on the management biology of North American black bear. *In* D. Burk, ed. The black bear in modern North America. The Amwell Press, N.J. 300pp.

30. LAWRENCE, W. 1979.
Proceedings of the workshop on the management biology of North American black bear. *In* D. Burk, ed. The black bear in modern North America. The Amwell Press, N.J. 300pp.

31. SHAFFER, S. 1971.
Some ecological relationships of grizzly bears and black bears of the Apgar Mountains in Glacier National Park, Montana. M.S. thesis, University of Montana, Missoula. 134pp.

32. MARTINKA, C. J. 1972.
Habitat relationships of grizzly bears in Glacier National Park. Prog. Rep. 1972, Natl. Park Serv., Glacier Natl. Park, Mont. 19pp.

33. MINORE, D. 1975.
Observations on rhizomes and roots on *Vaccinium membranaceum*. U.S. For. Serv. Res. Note PNW-261. 5pp.

34. MARTIN, P. 1979.
Productivity and taxonomy of the *Vaccinium globulare, V. membranaceum* complex in western Montana. M.S. thesis, University of Montana, Missoula. 136pp.

35. MUEGGLER, W. 1965.
Ecology of seral shrub communities in the cedar-hemlock zone of northern Idaho. Ecol. Monogr. 35:165–85.

36. FRANKLIN, J. F., AND C. T. DYRNESS. 1973.
Natural vegetation of Oregon and Washington. U.S. For. Serv. Gen. Tech. PNW-8. 417pp.

37. HEMMER, D. 1975.
Serviceberry: Ecology, distribution, and relationships to big game. Montana Fish and Game Dept., Helena. Job Compl. Rep. Proj. W-120-R-5 and 6. 76pp.

38. HAGAR, D. 1960.
The interrelationships of logging, birds, and timber regeneration in the Douglas-fir region of northwestern California. Ecology. 41:116–25.

39. AHLGREN, C. 1966.
Small mammals and reforestation following prescribed burning. J. For. 64:614–18.

40. ROGERS, L. L. 1976.
Effects of mast and berry crop failures on survival, growth, and reproductive success for black bears. Proc. North Am. Wildl. and Nat. Resour. Conf. 41:431–38.

41. MEALEY, S. P. 1977.
Method for determining grizzly bear habitat quality and estimating consequences of impacts on grizzly bear habitat quality. Final Draft. U.S. For. Serv. Reg. One. 36pp.

42. R. GREENE, IDAHO DEPT. LANDS, BOISE, PERS. COMMUN.

43. PFISTER, R., B. KOVALCHIK, S. ARNO, AND R. PRESBY. 1977.
Forest habitat types of Montana. U.S. For. Serv. Gen. Tech. Rep. INT-34. 174pp.

44. BEECHAM, J. 1982.
Black bear ecology. Idaho Dept. Fish and Game Job Prog. Rep., Fed. Aid Proj. W-160-R-6.

45. BARNES, V. G., AND O. E. BRAY. 1967.
Population characteristics and activities of black bears in Yellowstone National Park. Final Rep. Colorado Coop. Wildl. Res. Unit, Colorado State University, Ft. Collins. 199pp.

46. TISCH, E. L. 1961.
Seasonal food habits of the black bear in the Whitefish Range of northwestern Montana. M.S. thesis, Montana State University, Missoula. 108pp.

47. ZAGER, P. E. 1981.
Northern Selkirk Mountains grizzly bear habitat survey, 1981. U.S. For. Serv., Idaho Panhandle Natl. For. Contract. 75pp.

48. ZAGER, P. E., AND C. J. JONKEL. 1983.
Managing grizzly bear habitat in the northern Rocky Mountains. J. For. 81(8):524–26.

49. MARTIN, P. 1983.
Factors influencing globe huckleberry fruit production in northwestern Montana. Int. Conf. Bear Res. and Manage. 5:159–65.

ENDNOTES

[1] Bears used a habitat more often than researchers expected (for example, if 50 percent of an area was timbered, bears used that habitat more than 50 percent of the time).

[2] Bears used a habitat as often as researchers expected (for example, if 10 percent of an area was timbered, bears used that habitat about 10 percent of the time).

[3] Bears used a habitat less often than researchers expected (for example, if 50 percent of an area was timbered, bears used that habitat less than 50 percent of the time).

[4] Plant reproduction via underground stems or roots.

EIGHT | FOOD HABITS

IT ISN'T HARD FOR MOST PEOPLE WHO SPEND A good deal of time outdoors to visualize what it's like to be a wildlife research biologist. They can see themselves capturing and handling animals and collecting data on these animals. Few people realize, however, that biologists must also be good botanists; they must be able to describe the habitats that animals use and identify the plants that they eat.

Every field season, we scoured our study areas for bear scats, and analyzed the scats to find out which foods were important to the bears in those areas. To most of the field crew, the scats we collected each year were large, amorphous globs of who-knows-what. However, by the time Jan Brown, our food habits expert, finished with the scats, we knew not only what but also how much the bears were eating.

I've tried many times to describe Jan's work to those who want to know more about the feeding habits of black bears. Difficult and tedious are words that come to mind. Imagine going into your neighborhood and collecting a few flowers, leaves, and berries from a variety of shrubs, trees, and flowering plants. Put them in a blender and for fun, add an assortment of ants, bees, wasps, and beetles. Thoroughly grind this concoction and pour it into a shallow pan with some water. Now try to identify the kinds of plants and animals you collected and tell us how many of each kind there are. If you can imagine doing this work, I'm sure you can appreciate how hard it was to gather the information included in this chapter.

JOHN BEECHAM

The distribution and abundance of food affects the nutritional status of black bears and their growth rates, productivity, movement patterns, and survival. We need to know about bear food habits to help assess the quantity and quality of habitat available to bears, determine the impacts of land use decisions on bear habitat, and develop management programs to enhance or protect bear habitats.

Black bears are frequently described as opportunistic omnivores. Research on black bear food habits shows that bear feeding patterns are similar in different parts of the country – food habits in Idaho showed striking similarities to those reported for other areas in North America (1–7). Major differences occurred only in the species of plants and animals the bears ate.

The black bear's digestive system is shorter than that of most herbivores and longer than that of true carnivores. Because bears have a simple stomach and no *caecum*, they don't carry the types of microorganisms that plant eaters usually have to assist in digesting cellulose and other plant fibers. Black bears must eat large amounts of food each day to compensate for the low amounts of nutrients they extract from their mostly plant diet.

To learn what bears eat, we collected 2,057 scats during our studies of Idaho black bears. We separated these scats into two groups for analysis. During the spring and early summer, bears ate less nutritious foods such as grasses, broadleaf forbs, and horsetails (Plate 6). In late summer and fall, they ate a more nutritious diet dominated by the fruits of berry-producing shrubs (Plate 7). Bears frequently ate social insects (ants, bees, and wasps) throughout the active season.

SPRING AND EARLY SUMMER FOODS

In early spring, bears tended to feed on newly emerged plants that contained more digestible material and less fiber. We often saw bears feeding below the snow line after leaving their winter dens. As the season progressed, they followed the receding snow line to higher elevations until the first berries began ripening at lower elevations. Changes in mean elevation at which we found radio-collared bears supported our observations. In those years when early season berry crops were poor or failed, black bears continued to feed on green vegetation and insects until August (5).

Green vegetation was the most important component of the black bear's diet during this period. Grasses, forbs, and horsetails made up 62 to 94 percent of their diet.

Black bears at Priest Lake ate more horsetails and fewer forbs than bears at Council (Figs. 8–1 and 8–2). These data supported the observations of other researchers that bears used a large variety of foods in the spring, depending on their availability (1, 7, 8, 9).

Our trapping data showed that some bears were able to maintain their body weight during the spring and early summer, although most bears lost weight. Jonkel and Cowan (10) made the

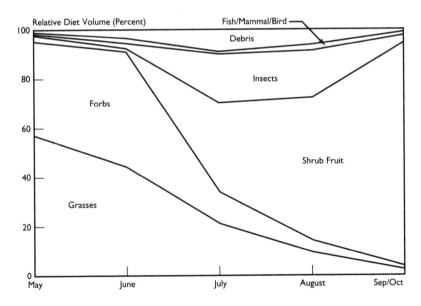

Figure 8–1. Relative percent volume of foods observed in the diet of black bears at Council, 1973–77.

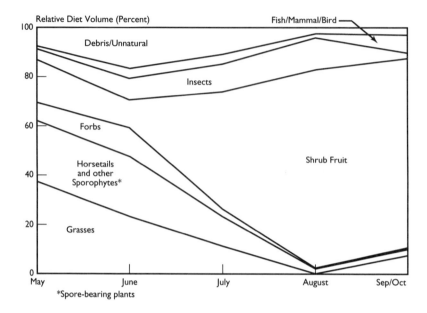

Figure 8–2. Relative percent volume of foods observed in the diet of black bears at Priest Lake, 1979–81.

same observation in Montana and referred to this time as a "negative foraging period."

LATE SUMMER AND FALL FOODS

Black bear food habits during this period reflected the change from a low-nutrition diet to a high-energy one. In the Council area, huckleberries and buffaloberries were ripening at middle elevations (4,500 to 5,000 feet) by mid-July, and Council bears concentrated their feeding activities at those elevations. In August, when supplies of those berries began to dwindle, bears began feeding on newly ripened bitter cherries, chokecherries, and hawthorn berries at lower elevations. Bears remained concentrated around these food resources until they denned (see chapter 6).

In northern Idaho, huckleberries were the most important shrubs used by bears (they used serviceberry and mountain ash in lesser amounts). Foraging patterns were similar to those we observed at

Council: Bears used low-elevation sites early and moved up as the berries ripened at higher elevations.

The most significant difference in food habits we noted between the Council and northern Idaho study areas was in the variety of berry species[1] bears used. Because the Council area supported a greater variety of berry-producing plants than did the northern Idaho areas, bears at Council used an average of eight berry species each year whereas those in northern Idaho averaged only three species.

This difference had major consequences for the two bear populations. Berries provided an abundant supply of energy-rich, digestible sugars essential to black bear reproduction and survival (2, 11, 3, 9). However, berry production varied from year to year, depending on spring temperatures and early summer precipitation levels, and dramatically affected reproduction and survival. The large variety of berry-producing shrubs used by Council bears made the bears less vulnerable to crop failures in one or more berry species. When huckleberries or buffaloberries were in short supply at Council, the bears fed extensively on ants during dry periods or continued feeding on green vegetation until the cherries and hawthorns ripened in mid-August. However, bears at Priest Lake, who used only a few berry-producing shrubs, were very vulnerable to huckleberry crop failures. In 1979, a major crop failure at Priest Lake resulted in decreased bear productivity and survival for two years.

The late summer/fall feeding period was obviously the most critical for black bears because of its influence on reproduction and the survival of subadults. In addition to this study, Jonkel and Cowan (10) and Rogers (2) observed reduced bear productivity and more deaths from starvation when major berry crop failures occurred. Increased black bear damage complaints were also reported when drought or late spring freezes damaged berry crops (2, 12, 13, 4, 5).

ANIMAL MATTER

Animal foods played a small but important role in black bear feeding habits in Idaho (insects were the most important animal food eaten). Bears sought out and ate large quantities of social insects, especially ants, during the transition period between early and late summer. In all months, we found ant eggs, larvae, and adults in bear scats. Ants played a particularly important role in the diet of black bears during drought years and when early season berry crops failed. We also saw bees and wasps frequently and beetles and flies occasionally in scats.

Another component of bear scats was vertebrate animals (mammals, fish, and birds). During all months, we found these foods in minor amounts (less than 2 percent). We used spawned-out salmon and steelhead as bear trap bait, and they probably accounted for most of the fish we found in the scats. The flesh, hair, and bones of mammals made up less than 1 percent of the black bears' diet, except in June and July at Council, when deer and elk remains accounted for 1.4 and 1.2 percent, respectively, of the bears' diet. Although any source of protein was important to bears in late spring when they were still losing body weight, we don't believe that the volume of deer and elk remains we found was sufficient to be considered critical to the bears. Many researchers studying food habits reported similar findings (14, 1, 10, 2, 11, 3, 12, 13, 6, 15). Bird remains turned up infrequently and in small amounts. We also found bear remains in scats – mostly hair that bears ingested while grooming themselves.

In addition, we found debris and other unnatural foods in scats every month. Most of these items were eaten by bears accidentally as they fed on other foods. Some of the things found in scats at Priest Lake were garbage items from nearby summer homes and camping areas.

MANAGEMENT IMPLICATIONS

Maehr and Brady (6) stated that habitat and plant diversity were important to black bears because they allowed bears to spend less energy searching for alternate foods when major foods were scarce. Our data supported this hypothesis and suggested that land management activities that affect habitat can help or harm bear populations.

Hammond (5) reported that livestock grazing on berry-producing shrubs reduced berry crops. Timber cutting opened the forest canopy, allowing sunlight to reach the forest floor and thus increase bear food production. Logging was especially helpful to bears if a few trees were left standing to provide escape cover. Plants sown along forest roads (clovers and dandelions) were important food sources. However, logging roads provided increased hunter access to bear habitats, which resulted in higher bear death rates and offset the beneficial value of some logging methods.

When making land use decisions, land managers need to take into account the impact of these decisions on bear habitat. The black bear is clearly an adaptable species capable of using many kinds of habitats in Idaho if we consider their needs in our management actions.

LITERATURE CITED

1. HATLER, D. F. 1972.
   Food habits of black bears in interior Alaska. Can. Field-Nat. 86:17–31.

2. ROGERS, L. L. 1976.
   Effects of mast and berry crop failures on survival, growth, and reproductive success of black bears. Trans. North Am. Wildl. and Nat. Resour. Conf. 41:431–38.

3. BEEMAN, L. E., AND M. R. PELTON. 1980.
   Seasonal foods and the feeding ecology of black bears in the Smoky Mountains. Int. Conf. Bear Res. and Manage. 4:141–48.

4. GRENFELL, W. E., AND A. J. BRODY. 1983.
   Seasonal foods of the black bears in Tahoe National Forest, California. California Fish and Game. 69:132–50.

5. HAMMOND, F. M. 1983.
   Food habits of black bears in the Greys River Drainage, Wyoming. M.S. thesis, University of Wyoming, Laramie. 50pp.

6. MAEHR, D. S., AND J. R. BRADY. 1984.
   Food habits of Florida black bears. J. Wildl. Manage. 48(1):230–35.

7. MAEHR, D. S., AND J. T. DEFAZIO, JR. 1985.
   Foods of black bears in Florida. Fla. Field-Nat. 13:8–12.

8. MACE, R. D., AND C. J. JONKEL. 1986.
Local food habits of the grizzly bear in Montana. Int. Conf. Bear Res. and Manage. 6:105–10.

9. HAMER, D., AND S. HERRERO. 1987.
Grizzly bear food and habitat in the front ranges of Banff National Park, Alberta. Int. Conf. Bear Res. and Manage. 7:199–213.

10. JONKEL, C. J., AND I. M. COWAN. 1971.
The black bear in the spruce-fir forest. Wildl. Monogr. No. 27. 55pp.

11. ROGERS, L. L. 1978.
Effects of food supply, predation, cannibalism, parasites, and other health problems on black bear populations. Bunnell, Eastman, and Peek, eds. Symp. Sat. Reg. Wildl. Populations. Vancouver, B.C.

12. GOLDSMITH, A., M. E. WALRAVEN, D. GRABER, AND M. WHITE. 1981.
Ecology of the black bear in Sequoia National Park. Natl. Park Serv. Final Rep. Contract No. CY-8000-4-0022. 64pp.

13. GRABER, D. M. 1982.
Ecology and management of black bears in Yosemite National Park. Final Rep. Natl. Park Serv., Yosemite Natl. Park. 206pp.

14. TISCH, E. L. 1961.
Seasonal food habits of the black bear in the Whitefish Range of northwestern Montana. M.S. thesis, Montana State University, Missoula. 108pp.

15. RAINE, R. M., AND J. L. KANSAS. 1990.
Black bear seasonal food habits and distribution by elevation in Banff National Park. Alberta Int. Conf. Bear Res. and Manage. 8:297–304.

ENDNOTES

¹ Each species made up more than 5 percent of the scat sample, by volume.

PLATE 1

To immobilize the snared bears, we injected
them using a syringe mounted on a jab stick.
Photo by Doyle Reynolds

PLATE 2
Color changes in a brown-phase bear
(NO. U–19). The upper photo was taken
in May 1973; the lower photo, in July
1973.   *Photo by John J. Beecham.*

PLATE 3
A female bear and her mixed-color-
phase litter. *Photo by John J. Beecham.*

PLATE 4

When we approached snared bears,
some of them tried to hide. We also
described five other types of bear
behavior at the trap site (see Chapter
6). *Photo by John J. Beecham.*

PLATE 5
The Council Mountain area contained
a variety of bear habitats, including timber,
open timber/shrubfield and meadow habitats.
Photo by John J. Beecham.

PLATE 6
Facing page. Forbs such as *Lomatium* form an important part of the bears' spring and early summer diet. *Photo by John J. Beecham*

PLATE 7
During late summer and fall, bears ate bitter cherries and other berries. These high-energy foods enabled bears to gain enough weight to survive winter. *Photo by Ron Beecham*

PLATE 8
Black bear cubs were
born in late January or
early February.
At birth, the bears
weighed eight to 12 ozs.
Photo by Doyle Reynolds

PLATE 9
Below. Most of the bear
dens we found were
ground dens, dug into
a hillside or beneath
trees or other vegetation.
Photo by John J. Beecham

NINE | POPULATION CHARACTERISTICS

WITHOUT QUESTION, THE TRAPPING PORTION OF our research provided the most excitement. It gave us the opportunity to actually get our hands on our subjects. Part of the excitement came from being close to a potentially dangerous animal and part came from the challenge of learning about an animal that few people see in its natural environment.

Black bears were relatively easy to catch the first time they visited a trap site. Catching them a second time, however, proved more difficult, and required ingenuity and persistence. Luckily the bears were somewhat predictable in their behavior, a fact that we used to our advantage. The bears often approached our trap sites from the same direction; they even put their feet down in the same spot each visit. All we had to do was figure out how the bears were coming into the trap site, and then we could usually find a way to catch them.

After a few years of trapping, we had developed a variety of techniques for catching trap-shy bears and were always trying to think of new methods. This led to a sense of camaraderie among the trappers and it soon became an us-versus-them competition. Checking our traps daily, rain or shine, was similar to having Christmas every day. Each time we approached a trap site, we felt excitement and anticipation. We never knew whether we had caught a particularly crafty bear or something more exotic in our snares. During our studies, we trapped mountain lions, coyotes, ravens, deer, elk, cows, dogs, and sometimes people. We even caught some of the research team in exceptionally well-camouflaged snares.

Our trapping days weren't always full of excitement, however. I remember 10-day trapping periods in late July and early August when temperatures soared to over 100°F. During those periods, we faithfully checked our snares each day, but were rarely rewarded with the sight of a trapped bear. Sometimes we caught as few as two or three bears in a 10-day period.

On the other hand, we also had some very productive days. In one day at Council, we captured 12 bears. On another day at Lowell, we captured 10 bears with 15 snares. We had mixed emotions about these productive periods, for they meant 16- to 18-hour workdays with little or no time to eat, not to mention no time to smell the flowers!

JOHN BEECHAM

Collecting information on population characteristics was one of the most important parts of our research because this information tells us whether bear populations are thriving or in trouble. We examined the age structure (percentage of older vs. younger bears) and sex composition (percentage of males vs. females) of each bear population. We also studied reproductive characteristics such as breeding season, age of first reproduction, and reproductive rates. Although it was a difficult task, we estimated population sizes and densities (the number of bears per square mile), and developed a technique to measure bear population trends. Finally, we looked at short- and long-term changes in population size (population regulation).

Gathering population data was sometimes hard. Black bears are secretive animals who live in forested habitats, so it's difficult to see them. Even where they are abundant, their numbers are low. We used *capture-mark-recapture* methods to obtain data on the population status of bears in six areas of the state. During our studies, we captured 707 black bears 1,396 times (Table 9–1). The number of

Table 9–1. Number of black bears captured in Idaho, 1978–89

Area	Captures Individuals	Total	Average Trap Days/Bear[b]	Age of Oldest Male Trapped	Age of Oldest Female Trapped
Council	278	625	8.1	22	24
Priest Lake	180	399	7.8	23	21
Lowell	156	237	6.0	21	18
CDA River	62	96	37.7[a]	19	15
St. Joe River	21	25	37.6	13	19
Elk River	10	14	60.0	10	8
Total	707	1,396	–	–	–

[a] Bears were not equally distributed on study area: 26 of 37 bears captured in 1978 were caught in the unlogged half of the study area, although each half had an equal number of trap sites. As a result, this figure is inflated.
[b] The average number of trap days (one trap set for one day) required to capture a bear.

males and females recaptured varied considerably from year to year, but most bears were recaptured once or twice each year. However, we snared one male 7 times during a single summer and caught another male at Council 11 times between 1973 and 1982. The oldest male captured was 23 (Priest Lake); the oldest female was 24 (Council).

To make it easier to compare the Council, Priest Lake, and Lowell bear populations, we calculated a minimum population size for each area (the *reconstructed population*) by adding the number of adults and subadults (bears less than four years old) captured the first year to the number of unmarked bears captured in subsequent years that were alive the first year. This gave us a reconstructed population for the first year. These data indicated that we caught 81 to 100 percent of the adult males and 78 to 94 percent of the adult females by the end of the second trapping season in each study area. Although the reconstructed population represented only the minimum number of bears present in the first year, it was useful in making comparisons among the three study areas.

AGE STRUCTURE

To figure out the age structure of a bear population, we calculated its median age and the proportion of nonbreeding or subadult bears it contained. In general, healthy bear populations are dominated by older bears, as reflected in higher median ages and lower proportions of nonbreeding bears.

COUNCIL

We captured 278 bears (152 males; 126 females) 625 times with snares during 1973–1977, 1982, and 1988–1989. The oldest male we caught was 22; the oldest female, 24.

The Council bear population consisted of many young animals. The median age for the captured animals varied from three to four years (four for the reconstructed population). The proportion of nonbreeding bears ranged from 38 to 67 percent (Table 9–2) and was 31 percent for the reconstructed population.

Table 9–2. Age structure of black bears captured at Council, 1973–89

| | Number of Males (M) and Females (F) Captured Each Year | | | | | | | | | | | | | | | |
| | 1973 | | 1974 | | 1975 | | 1976 | | 1977 | | 1982 | | 1988 | | 1989 | |
Age Class	M	F	M	F	M	F	M	F	M	F	M	F	M	F	M	F
Cubs	3	2	1	0	7	4	0	1	8	7	1	0	0	0	0	1
Yearlings	4	3	1	1	2	1	10	5	4	3	4	3	2	4	11	5
Subadults	9	7	11	2	12	4	9	5	16	17	7	3	3	6	5	10
Adults	13	15	12	14	15	6	8	19	13	14	15	12	7	13	10	18
Nonbreeders %	50		38		59		53		67		40		43		53	

During our 1973–1977 studies, we captured more 2.5-year-old males than 2.5-year-old females, but we recaptured the males less frequently. In 1988 and 1989, we didn't see this pattern of higher capture rates for 2.5-year-old males. These data suggested that during 1973–1977, 2.5-year-old males frequently moved through our study area. The radio-telemetry data we collected on subadult movement patterns showed they were very mobile and supported this observation. The fact that we didn't see a significant influx of 2.5-year-old males into the study area in 1988 or 1989 suggested that fewer young bears were being produced in areas adjacent to the Council study area than during our 1973–1977 studies.

We found no differences in recapture rates between 3.5-year-old males and females, adult males and females, or subadult and adult bears.

PRIEST LAKE

We captured 180 bears (123 males; 57 females) 399 times during our Priest Lake studies in 1979–1981 and 1988. The oldest male captured was 23; the oldest female, 21.

The Priest Lake population contained more older bears than did the Council population, the median age ranging from six to seven years (five for the reconstructed population). The proportion of nonbreeding bears at Priest Lake (22 percent for the reconstructed population) wasn't significantly different from the value we observed at Council, but it was lower than the Council value and higher than the Lowell value (Tables 9–3 and 9–4).

According to our capture-recapture data, we caught more 2.5- and 3.5-year-old males than females. Because of small sample sizes for these age groups, however, we weren't able to show that we recaptured the males less frequently. We saw no differences between

Table 9–3. Age structure of black bears captured at Priest Lake, 1979–88

| | Number of Males (M) and Females (F) Captured Each Year | | | | | | | |
| | 1979 | | 1980 | | 1981 | | 1988 | |
Age Class	M	F	M	F	M	F	M	F
Cubs	5	2	2	0	1	0	0	0
Yearlings	5	4	4	1	0	0	0	0
Subadults	17	3	9	2	15	3	6	3
Adults	33	17	23	19	22	11	31	5
Nonbreeders (%)	42		30		37		20	

Table 9–4. Age structure of black bears captured at Lowell, 1975–79

	Number of Males (M) and Females (F) Captured Each Year							
	Preremoval Years[a]				Postremoval Years			
	1975		*1976*		*1977*		*1979*	
Age Class	M	F	M	F	M	F	M	F
Cubs	0	0	4	3	0	0	1	0
Yearlings	3	0	1	2	0	0	1	0
Subadults	5	2	7	3	2	4	5	5
Adults	19	12	29	26	17	12	18	23
Nonbreeders (%)	24		27		17		23	

[a] Bears captured in 1976 were moved to other areas of Idaho as part of a calf elk study.

the recapture rates of adult males and females or subadults and adults.

The large movement of nonbreeding males into the study area was similar to that seen at Council. It's likely the young males were moving through the area after dispersing from adjacent habitats.

Subadult movements into and out of the Council and Priest Lake bear populations were partly responsible for the higher proportion of subadults in these populations and helped maintain population sizes. Kemp (1, 2) reported a similar increase in the number of subadult males on his Cold Lake, Alberta, study area after 26 adult males were removed from the population. He suggested that this influx of subadults caused the population to double in two years. However, our data indicated that many of the subadults (primarily males) who moved into our study areas were transients who didn't remain in the area. Thus, no real increase in population size occurred.

LOWELL

We captured 156 bears (83 males; 73 females) 237 times on the Lowell study area. We caught 225 bears with foot snares and darted 12 bears from a Hiller 12-E helicopter. The oldest male we caught in this area was 21; the oldest female, 18.

The Lowell bear population contained a significantly larger proportion of adults (84 percent) than either the Council or Priest Lake populations; the median age was seven. The proportion of nonbreeding bears ranged from 17 to 27 percent and was 16 percent for the reconstructed population (Table 9–4).

It was impossible to analyze capture-recapture data from Lowell because of small sample sizes in the subadult age groups and because all bears captured in 1976 were moved to other areas of the state as part of an ongoing calf elk survival study. However, we anticipated a large influx of 2.5- and 3.5-year-old males into the study area in 1977 and 1979 to fill habitat left vacant because of our translocation effort in 1976. We expected to see this increase because we observed these kinds of movements at Council and Priest Lake. In addition, the Lowell study area is bordered by a large wilderness area where hunting pressure is low and older bears dominate. We believed young males from the wilderness area might disperse to the adjacent vacant habitats we created at Lowell in 1976.

However, we saw no increase in 2.5- or 3.5-year-old males in 1977 or 1979. (We captured two young males in 1977 and five in 1979.) Although the bear population declined in 1977, it appeared to have recovered substantially by 1979. The average number of days needed to trap a bear went from 6.2 in 1976 to 7.5 in 1977 to 4.4 in 1979. We weren't able to detect a significant increase in any age class, and the median age didn't drop appreciably – it went from seven to six years.

COEUR D'ALENE RIVER

We captured 62 bears (41 males; 21 females) 96 times during two field seasons (1978 and 1983) at Coeur d'Alene. The oldest male captured was 19; the oldest female was 15.

Although our studies at Coeur d'Alene were single-season status surveys and didn't yield the same quantities of data that we collected at Council, Priest Lake, and Lowell, we did observe some

Table 9–5. Age structure of black bears captured at the Coeur d'Alene River area in 1978 and 1983

| | Number of Males (M) and Females (F) Captured Each Year | | | |
| | 1978 | | 1983 | |
Age Class	M	F	M	F
Cubs	1	1	3	1
Yearlings	3	0	0	0
Subadults	2	3	8	4
Adults	19	8	5	6
Nonbreeders %	27		59	

interesting changes between 1978 and 1983 (Table 9–5). Before 1978, this study area was managed conservatively; bear harvest was low during a fall hunting season. After our initial study in 1978, a spring bear season was added. We designed our 1983 status survey to evaluate the impact of the longer season on the bear population.

Our capture data from these two surveys revealed that the proportion of nonbreeding bears in the Coeur d'Alene River population increased significantly – from 27 percent in 1978 to 59 percent in 1983. The median age of the population declined from seven to two years. The median age for captured males dropped from eight to two years, while the median age for females increased from five to six years. In 1978, the oldest male we captured was 19; the oldest female, 15. In 1983, the oldest male we captured was nine; the oldest female, 12.

These data supported data collected from other study areas in Idaho that showed greater hunting pressure caused numbers of subadult males to increase and numbers of adult males to decrease in bear populations (Fig. 9–1). Our data provided dramatic evidence that adult males are particularly vulnerable to hunting and disappear quickly when hunting pressure is increased.

In the past 20 years, researchers have conducted many studies on black bear populations that were hunted – some lightly, others

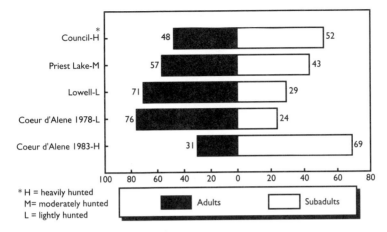

Figure 9–1. Proportions of adult males captured in heavily hunted and lightly hunted areas.

quite heavily (3–6). These studies and others on grizzly/brown bears showed that bear populations had similar responses to increasing harvest pressure (7–9). Lightly hunted populations were dominated by older bears: The median age was at least six years, more than 60 percent of the males were adults, and more than 75 percent of the females were adults (Table 9–6). In heavily hunted populations, males made up less than 50 percent of the total population, and they were younger than females (6, 9).

Table 9–6. Trends observed in lightly hunted and heavily hunted black bear populations

Lightly Hunted	Heavily Hunted
Proportion of adult males >60%	Males <50% of total population
Proportion of adult females >75%	Median age of females higher than males
Median age >6	Median age <4
Proportion of nonbreeders <30%	Proportion of nonbreeders >45%

Kolenosky (6) reported that adult males were the most vulnerable segment of a bear population to hunters; subadult males were somewhat less vulnerable. All marked males on his study were shot within six years of tagging whereas marked females were still being harvested as long as 11 years after being tagged. Kolenosky also stated that three- to six-year-old females were more vulnerable to hunters than older females. Willey (10) reported comparable results for black bears in Vermont.

We observed similar age and vulnerability relationships. In the lightly hunted Lowell and 1978 Coeur d'Alene River populations, 71 and 76 percent of the males were adults, and their median ages were high – 7.5 years (Fig. 9–1). The Council and 1983 Coeur d'Alene River populations contained fewer adult males (48 and 31 percent), and their median ages were lower – 4.5 and 2.5 years. These data suggested that the Council and 1983 Coeur d'Alene River bear populations were heavily hunted (3, 7). The Priest Lake population fell between the lightly and heavily hunted populations: It contained 57 percent adult males, and its median age was 5.5 years.

SEX COMPOSITION

Sex composition data collected from capture-recapture studies are usually biased and hard to interpret because capture methods and bear behavioral differences, nutritional condition, and reproductive status can all affect how vulnerable black bears are to capture. Their large home range and aggressive behavior often make males easy to catch. Poor nutritional condition and the presence of cubs can make females either very easy or difficult to catch (8).

Data from long-term (greater than two years) trapping studies should show more adult males than females in lightly or unhunted populations (11, 6). Heavily hunted populations should contain more females (12). We captured more males on all study areas in all years except during three years at Council (1986, 1988, and 1989). However, only in 1988 did we capture significantly more females than males at Council (Table 9–7).

To determine the sex composition of a bear population, we

Table 9–7. Age structure of reconstructed black bear populations at Council in 1973 and 1988

	Number of Males (M) and Females (F) Captured Each Year			
	1973		*1988*	
Age Class	M	F	M	F
Cubs	3	2	11	5
Yearlings	4	3	2	4
Subadults	9	7	3	6
Adults	15	21	13	27[a]
Nonbreeders (%)	44		44	

[a] Significantly different from a 50:50 ratio.

kept track of the number of males and females captured in the 2.5, 3.5, and adult age groups. To get a truer picture of male:female ratios, we decided to use data from our reconstructed populations to help us identify differences in sex composition among these populations.

COUNCIL

In most years of the study, we caught more adult and subadult males than females. The *sex ratio* of the reconstructed population also favored males. However, in 1986, 1988, and 1989 we trapped more females than males.

During the study, we saw an interesting difference between the proportion of subadult males present in 1973–1977 and that present in 1988–1989. In 1973–1977, we caught more 2.5-year-old males than females. Jonkel and Cowan (13), Kemp (1, 2), and Beecham (4, 5) reported that large numbers of subadult males moved into and through hunted bear populations in Montana, Alberta, and Idaho. Rogers (12), however, didn't see this phenomenon in the hunted population he studied in Minnesota, although all males born on his study areas dispersed as 2.5- or 3.5-year-olds. In 1988 and 1989, the same thing happened at Council: We caught more 2.5-year-old females than males and we saw no influx of subadult males. Why the difference? During 1973–1977, lightly hunted bear populations existed near the Council and the Montana and Alberta areas. Subadult males moved from these lightly hunted or "reservoir" areas into the more heavily hunted areas, where vacant habitat was available. In areas where black bear hunting pressure was more evenly distributed across the bears' habitat, as it probably was in Minnesota and at Council during 1988–1989, there were fewer dispersing subadult bears. We concluded that smaller bear populations could be anticipated in heavily hunted areas unless there were reservoir areas nearby that could produce the subadults needed to repopulate heavily hunted areas. Miller (9) reported that males constituted a smaller proportion of the population in heavily hunted areas.

Council was the only study area where we captured significantly more females than males. We believe that the sex ratios we saw at Council during 1988 and 1989 meant that the population was heavily hunted. Because of the increasing number of females and the lack of subadult movement into the area during those years, we were concerned about the status of this bear population and recommended more conservative hunting seasons.

PRIEST LAKE

Analysis of our capture records shows that we trapped significantly more males than females in the 2.5, 3.5, and adult age groups. The reconstructed population for the Priest Lake area also was dominated by males.

Hunting pressure in the Priest Lake area was lighter than that observed at Council. In most years, the Priest River drainage was closed during the spring to bear hunting, and although road access was reasonably good, the area wasn't located near a major population center. The sex ratios we found at Priest Lake supported Kolenosky's (6) and Miller's (9) observations that males were more vulnerable than females to hunting. Although males dominated this population, they occurred in significantly lower numbers than we observed in lightly hunted populations at the Lowell and the

Coeur d'Alene River areas in 1978. The proportion of adult males in the Priest Lake bear population was 57 percent, compared with the 71 percent found at Lowell and the 76 percent at Coeur d'Alene River in 1978. The proportion of adult males at Priest Lake didn't differ significantly from that of the Council population in 1973–1977 (48 percent, see Table 9–8).

Table 9–8. Proportions of adult males and females captured at Council, Priest Lake, Lowell, and the Coeur d'Alene River

	Proportions of Adults Captured	
Area	Males (%)	Females (%)
Council, 1973–77 (Heavy Hunting Pressure)	48	36
Priest Lake (Moderate Hunting Pressure)	57	76
Lowell (Light Hunting Pressure)	71	79
Coeur d'Alene, 1978 (Light Hunting Pressure)	76	67
Coeur d'Alene, 1983 (Heavy Hunting Pressure)	31	55

We saw no significant differences between the proportion of adult females captured at Lowell (79 percent) and Coeur d'Alene River in 1978 (67 percent) and that of adult females captured at Priest Lake (76 percent). However, we caught significantly more adult females at Priest Lake (76 percent) than at Council (36 percent) in 1973–1977.

LOWELL

The sex ratios we saw at Lowell were similar to those described by Kolenosky (6) and Miller (9) for lightly hunted bear populations, except that we didn't capture more adult males than females. We believed this finding was influenced by the method we used to check our traps each day. Unlike the other studies we conducted, we checked all snares set on the Lowell area daily by helicopter. If the trap site wasn't disturbed, we didn't land, and flew on to the next trap site. This method probably reduced the amount of human scent around our sites and increased the proportion of females we captured.

Another interesting result of our capture data was caused by the distribution of our trap sites in the study area and the length of the trapping season (six weeks). We trapped this area for a shorter period, and we didn't saturate the area with trap sites as we did on other study areas. The sites were scattered throughout the area on ridges where we could land the helicopter. As a result, we captured only 45 and 27 percent, respectively, of the adult males and females the first year, and 81 and 78 percent the second year.

Although the overall sex ratio of black bears captured on the Lowell study area was 50:50, we did catch significantly more 2.5-year-old males than females during 1975 and 1976. After we removed 75 bears in 1976, we captured equal numbers of males and females in all age groups.

We trapped more adult males at Lowell than on any other study area except the Coeur d'Alene River area in 1978. We caught more adult females at Lowell than we did at Council during 1973–1977, but not significantly more than on other study areas in Idaho (Table 9–8). These data further support our belief that this population was lightly hunted.

COEUR D'ALENE RIVER

Analysis of our 1978 Coeur d'Alene River capture data suggested that this population was lightly hunted. We trapped significantly more males than females, and the proportion of adult males in the population (76 percent) was high.

As a result of this survey, the black bear hunting season was liberalized by opening a spring season. Five years later we returned to this study area to resample the bear population and evaluate the impacts of the longer hunting season. Our capture data from this survey showed significantly fewer adult males in 1983 (31 percent) than in 1978 (76 percent). We detected no difference in the number of males and females captured.

These data supported the observations of others that adult males were more vulnerable to hunting than adult females (11, 12, 6, 9). They were also consistent with our findings from other study areas in Idaho.

REPRODUCTION

A black bear population is limited in how fast it can grow or recover from a population decline by its reproductive characteristics: how much time is available for breeding (the breeding season), how old females are when they first produce young (the age of first reproduction), and how fast females in a population can produce cubs (the population's productivity).

Black bears are a classic K-selected species. That is, they are long-lived, they mature late, and they have low reproductive rates. These characteristics made it hard to sample enough bears from each study area to allow valid comparisons among study areas. As a result, we have combined data from the Council, Priest Lake, Lowell, and Coeur d'Alene River areas to evaluate the reproductive biology and productivity of black bears in Idaho.

BREEDING SEASON

The black bear breeding season extends from mid-May until early August in Idaho. During that time, males periodically visit females for intervals of 15 minutes to 2 hours to see if the females are ready to breed (14). If a female is in estrus, the male will remain with her for two to five days (14). The bears may breed several times during this period, and each breeding act may last as long as 15 minutes or more (15).

Ovulation (egg release) in female black bears is stimulated by the breeding act (16). After the egg is fertilized, the resulting embryo(s) implants in the female's uterine wall. The embryo doesn't begin growing until early December, however, and then it develops rapidly for six to ten weeks (17). Although the full gestation period is about 220 days, active embryo growth occurs for only 40 to 70 days.

To measure bear population trends, we used bait station surveys. We placed sardine cans in trees and counted the number of bear visits to these bait stations. Photo by Jeff Rohlman

Cubs are born in late January or early February, and are very dependent on their mothers at birth. They are born with closed eyes, have hair, and weigh 8 to 12 ounces (Plate 8).

To determine the breeding season for female black bears in Idaho, we used vaginal smears and the degree of swelling in the vulva. Although we were the first to use vaginal smears on black bears, we believed they were an objective method for determining the breeding season because of their similarity to vaginal smears from other carnivores (18–21).

Vaginal smears from 14 females showed ten bears were in estrus in June, three in the first half of July, and one in late July (22). These dates coincided closely with those obtained by rating the degree of vulva swelling and were similar to those reported by other researchers (16, 13, 23, 24, 8, 14).

We determined male bears' breeding status by measuring the length and width of their testes and calculating an equivalent diameter for each individual captured (ED = length + width/2). In bears older than four years, we found no significant differences in testes size for bears during the breeding season. These data suggested male black bears were reproductively active throughout the breeding season; they didn't peak in June as females did.

BREEDING AGE AND PRODUCTIVITY

We were able to determine when 27 females first produced young: 10 females (37 percent) first successfully conceived at 3.5 years, 11 (41 percent) at 4.5 years, and 6 (22 percent) at 6.5 years. The average age was 4.9 years. We saw no bears younger than 4 years or older than 18 years with cubs. In the Council area, all adult females 6 years or older had produced cubs. One of four 6-year-olds at Lowell and three of eight 6-year-olds at Priest Lake hadn't produced their first litter of cubs. On the Coeur d'Alene River area, one of three 7-year-olds hadn't yet produced her first litter.

These data showed that black bears in Idaho had lower ages of first reproduction than those in Montana (13), Alaska (25), or Ontario (6), but not as low as those reported for the eastern United States (26–30).

The average size of the 82 litters we observed was 1.7 cubs (Table 9–9), and the average litter frequency was 18 percent (range 0 to 62 percent). These data represent minimum values because we didn't capture all lactating females each year. In fact, capture and

Table 9–9. Minimum observed litter sizes for black bears in Idaho

Area	Number of Litters Observed	Number of Litters with:			Average Litter Size
		1 cub	2 cubs	3 cubs	
Council	32	7	23	2	1.8
Priest Lake	20	11	8	1	1.5
Lowell	23	10	11	2	1.7
Coeur d'Alene	7	2	5	–	1.3
Total	82	30	47	5	1.7

telemetry records suggested that because lactating females were difficult to catch, we probably captured very few of the females with cubs each year. Jonkel and Cowan (13) observed a similar pattern in Montana: The average litter frequency was 16 percent, but it ranged from 0 to 40 percent over an eight-year period.

Most of our radio-collared females bred every other year. However, this pattern wasn't consistent: Our capture records revealed that several bears were without cubs for at least two consecutive years. Although our capture records showed low and high years of cub production, differences in ages of first reproduction and periodic berry crop failures prevented bears from establishing the same breeding pattern (31). Kohn (32) reported that 31 percent of the adult females he captured in Wisconsin didn't breed for two years in a row and 8 percent skipped three years.

The age of first reproduction, litter size, litter frequency, and cub survival affect reproductive rates in black bear populations (6). However, regional and yearly differences in reproductive rates also occur. These variations may be primarily due to diet and nutrition (33, 34, 16, 13, 35, 36, 22, 6).

Rogers (36) reported that captive black bears on a rich diet developed more rapidly than wild bears and often bred at younger ages. In general, accumulated data on black bear reproductive biology indicate an east-west gradient in bear productivity associated with nutrition (22). In the East, where a hard mast (nuts) diet is generally available in spring and fall, bears have a lower age of first reproduction, higher reproductive rate, and larger average adult body weight than bears in the West, where a less-nutritious diet of grasses, forbs, and soft mast (berries) is most common (33, 34, 16, 27, 28, 37, 13, 35, 22).

Jonkel and Cowan (13) and Rogers (36) noted increased reproductive success after years when fall foods were abundant and decreased success when most crops failed. We observed a dramatic illustration of this type of response at Priest Lake in 1980. The 1979 fall huckleberry crop was one of the poorest local residents could remember. In 1980, none of the 20 females we caught were lactating, and in 1981, only 1 of the 11 adult females we captured had cubs.

Different population densities and/or age structures may also influence reproductive rates and the age of first reproduction. Kolenosky (6) reported that 5- to 7-year-old females produced smaller litters than females aged 8 to 18. We also observed smaller litters in young females and larger litters in females who were producing their second or third litters. Although nutrition appeared to account for most of the variability we saw in productivity, factors such as female longevity, large body size, possession of a home range, and protection from hunters were also important (22, 6).

POPULATION SIZE AND DENSITY

Estimating the size of populations was one of the hardest tasks we faced during our studies. Bears are difficult animals to see and count because they prefer timbered habitats. Even where black bears are abundant, they lead solitary lives and seldom associate with other bears. Density (the number of animals per unit area, i.e., square mile) is a difficult concept to apply to field studies of large, free-ranging mammals (38). Its usefulness for making comparisons between areas was limited because of differences in topography, land use patterns, bear harvest rates, bear behavior, and the methods we used to estimate density.

All of our study areas were small parts of bear habitats containing many square miles. We used each area as the basis for calculating bear densities. As a result, our density figures more accurately represented the number of bears whose home ranges overlapped the trapped area rather than the number of bears residing within the area (38). This technique overestimated the number of males present on each area and gave us higher density estimates than actually occurred because male black bears had larger home ranges than females (39, 22, 40).

We used the population reconstruction method and the Jolly-Seber method to calculate population sizes and densities. The Jolly-Seber method had two major assumptions: (1) no bear tags were lost, and (2) the chance of capturing any individual was the same for all individuals in the population. We met the first assumption by permanently marking every captured bear with a lip tattoo. However, we failed to satisfy the second assumption due to bear behavioral differences. Even so, the Jolly-Seber and population reconstruction methods produced similar estimates of population size.

From the population reconstruction method, we estimated that 77 black bears (1.5 bears/square mile) used the Council study area; 94 bears (0.8 bears/square mile) used the Priest Lake study area; and 116 bears (1.2 bears/square mile) used the Lowell study area. The Jolly-Seber method provided a second estimate of 98 bears (2.0 bears/square mile) at Council, 108 bears (0.9 bears/square mile) at Priest Lake, and about 55 bears (1.1 bears/square mile) on the Coeur d'Alene River area (Table 9–10). We couldn't use either method to estimate population sizes on the St. Joe River or Elk River areas because we didn't trap enough bears there. However, we did record the average number of trap days (one trap set for one day) required to capture a bear on each study area (Table 9–10). Although these data didn't provide a population estimate, they did serve as an index to population size.

Our bear densities were higher than those reported in Montana (13), Alberta (1, 2), Arizona (41), and Minnesota (12). Erickson and Petrides (42) reported a much lower density in Michigan.

POPULATION MONITORING

Because trapping was an expensive and time-consuming process, it wasn't feasible for monitoring the long-term status of bear populations. In 1983, we tested a bait station survey technique first devel-

Table 9–10. Population estimates and indices for six black bear populations in Idaho

Area	Area Trapped (mi^2)	Population Estimate		Mean Trap Days/Bear[a]
		Reconstructed Population Method	Jolly-Seber Method	
Council (1973–77)	50	77	98.5	6.5
Council (1988–89)	50	77	–	9.7
Priest Lake (1979–81)	115	94	108	7.8
Lowell (1975–79)	100	116	–	6.0
CDA River (1978)	50	–	57	28.4[b]
CDA River (1983)	50	–	53	47.0
St. Joe River (1982)	–	–	–	37.6
Elk River (1981)	24	–	–	60.0

[a] The average number of trap days (one trap set for one day) required to capture a bear.

[b] Bears were not equally distributed on the study area: 26 of 37 bears captured were caught on the unlogged half of the study area, although each half had an equal number of trap sites.

oped in Tennessee to see if it could effectively monitor bear populations. The technique was designed to measure population trends by counting the number of bear visits to a series of bait stations placed in known bear habitats. We used the Council, Priest Lake, Coeur d'Alene River, and St. Joe River areas as study sites.

Initial results from our bait station surveys correlated well with the trapping data we had collected at the Council, Priest Lake,

Coeur d'Alene River, and St. Joe River areas. Similar results were reported for bait station surveys in Tennessee (43) and Wisconsin (32). In 1988, we took our research one step further and removed bears from the Council and Priest Lake areas to test the sensitivity of this technique to a change in population size. At Council, spring and fall controlled hunts were established in 1988, 1989, and 1990 in an attempt to remove 25 to 30 bears from the population each year. Seventy-five permits were issued for each hunt. Hunters took 19 bears in 1988, 33 bears in 1989, and 20 bears in 1990. These harvest numbers met or exceeded our goal, and we observed a significant decline in the number of bear visits to the bait stations in 1990 (Fig. 9–2).

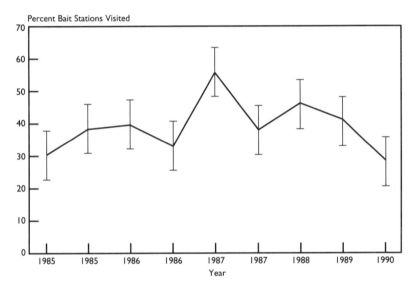

Figure 9–2. Percentage of bait stations visited at Council, 1985–90.

At Priest Lake in June 1988, we reduced the population by simply trapping 35 bears and moving them to other areas in Idaho. We ran our bait station surveys during the first week of July, before these bears were able to return to Priest Lake. We saw a significant decline in bear visits to our bait stations in 1988, but by 1989 visitation rates had returned to preremoval levels (Fig. 9–3). These data

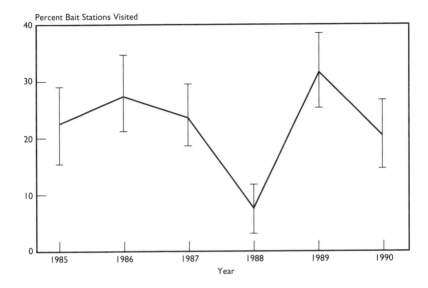

Figure 9–3. Percentage of bait stations visited at Priest Lake, 1985–90.

suggested that the bait station survey may be a useful tool for monitoring population trends that will enable managers to detect population changes of + 10 percent.

POPULATION REGULATION

Short-term or annual changes in the size of black bear populations
were related primarily to variations in birth rates, which were associated with the availability of nutritious foods. Jonkel and Cowan
(13), Rogers (12), and Reynolds and Beecham (22) reported that
berry crop failures apparently prevented adult females from producing young the following year. After one or more years of poor
berry production, Jonkel and Cowan (13) and Rogers (12) captured
starving bears, primarily subadults. At our Priest Lake study area,
none of the 20 adult females we captured in 1980 had produced
young after a berry crop failure in 1979. In 1981, only 1 of 11 adult females we captured was lactating, even though the 1980 berry crop
was above average.

Habitat quality and quantity and its distribution are the most important factors in long-term population control, because they influence bear reproduction. Rausch (25), Jonkel and Cowan (13), Rogers (36), and Reynolds and Beecham (22) found that bears receiving nutritionally superior diets matured earlier, had larger litters, and bred more often than did bears on poorer diets. Rogers (12) reported that female black bears had flexible territories that shifted according to the distribution and abundance of food, but he wasn't able to document that females actively and consistently defended these territories from unrelated bears. Amstrup and Beecham (39) and Reynolds and Beecham (22) reported that black bears in southern Idaho weren't territorial, but used home ranges that shifted seasonally and yearly in response to the availability of preferred foods. Rohlman (40) reported that black bears in northern Idaho used relatively stable home ranges, but didn't defend them. These differences may be related to the variability of food supplies in time and location and the abilities of bears to defend those food resources (44). The data further suggested that black bear social and reproductive systems have evolved that are regulated by the quality, quantity, and distribution of available food.

The dispersal of subadult bears may also affect population sizes. Kemp (1, 2) suggested that aggressive behavior by adult males toward subadult males caused the younger males to disperse, and he believed this behavior was an important regulating factor in bear populations. Rogers (12), however, working in an area with few adult males, found that nearly all subadult males born in his study area also dispersed as 2.5- and 3.5-year-olds. We, too, found that 1- and 2-year-old males left our study areas even when there were few adult males. At Council, all of the eight radio-collared male cubs we observed for three years left the study area.

Analysis of our growth and reproductive data indicated that although most males became sexually mature at 3.5 to 4.5 years, they didn't reach physical maturity (their maximum size) until they were 7.5 years old (4). Thus it may be hard for young males to compete with older males for females. Young males may have more breeding success by seeking out females on new ranges rather than waiting on their *natal ranges* until they become large enough to compete for females with older males. Barber and Lindzey (14) saw instances of dominant males displacing other males (presumably smaller, but at least subdominant) in courtship activities on their Long Island, Washington, study area.

Subadult females probably don't compete for mates as much as males do and, therefore, would gain little by dispersing into unfamiliar ranges. Lindzey (24) and Rogers (12) observed that subadult females remained in all or a portion of their mother's home range. Theoretically, however, the mother's home range can't support unlimited numbers of female offspring, and at high densities subadult females may also disperse or die. We observed an influx of both males and females into the Lowell study area after the removal of 75 bears in 1976. Kolenosky (6) also reported that he captured transient females on his Ontario study area. In high-density bear populations like those in the Selway-Bitterroot Wilderness next to the Lowell study area, adult females may play a role in prompting subadult females to disperse.

Subadult dispersal doesn't regulate population size unless the dispersing bears die. In fact, natural mortality rates may be lower in hunted populations, where space is available to accommodate dispersing bears, than in high-density populations where subadults are forced to enter marginal or less secure habitats. Hunted areas won't be repopulated unless they are near lightly hunted or unhunted reservoir areas that produce surplus subadults. In areas where heavy hunting pressure is evenly distributed across all bear habitats, population declines may occur. This situation is particularly serious in isolated habitats that lack sufficient access to reservoir areas.

LITERATURE CITED

1. KEMP, G. A. 1972.
   Black bear population dynamics at Cold Lake, Alberta. 1968–1970. Pages 26–31. In S. Herrero, ed. Bears – Their biology and management. Int. Union Conserv. Nat. New Ser. 23. Morges, Switzerland.

2. KEMP, G. A. 1976.

The dynamics and regulation of black bear, *Ursus americanus*, populations in northern Alberta. Pages 191–197. *In* M. R. Pelton, J. W. Lentfer, and G. E. Folk, eds. Bears – Their biology and management. Int. Union Conserv. Nat. New Ser. 40. Morges, Switzerland.

3. MCILROY, C. W. 1972.

Effects of hunting on black bears in Prince William Sound. J. Wildl. Manage. 36:828–37.

4. BEECHAM, J. 1980.

Population characteristics, denning, and growth patterns of black bears in Idaho. Ph.D. thesis, University of Montana, Missoula. 101pp.

5. BEECHAM, J. 1983.

Population characteristics of black bears in west central Idaho. J. Wildl. Manage. 47(2):402–12.

6. KOLENOSKY, G. B. 1986.

The effects of hunting on an Ontario black bear population. Int. Conf. Bear Res. and Manage. 6:45–55.

7. STRINGHAM, S. F. 1980.

Possible impacts of hunting on the grizzly/brown bear, a threatened species. Pages 337–49. *In* C. J. Martinka and K. L. McArthur, eds. Bears – Their biology and management. U.S. Gov. Printing Off., Washington, D.C.

8. BUNNELL, F. L., AND D. E. N. TAIT. 1981.

Population dynamics of bears – Implications. Pages 75–98. *In* C. W. Fowler and T. D. Smith, eds. Dynamics of large mammal populations. John Wiley and Sons, Ltd., New York, N.Y. 477pp.

9. MILLER, S. D. 1987.

Susitna hydroelectric project. Final Rep. Big game studies: Black bear and brown bear. Alaska Dept. Fish and Game. 276pp.

10. WILLEY, C. H. 1972.

Vulnerability of bears to hunting. Pages 24–27. *In* R. L. Miller, ed. Proceedings of the 1972 black bear conference. N.Y. State Dept. Environ. Conserv. Delmar, N.Y.

11. PEARSON, A. M. 1976.
Population characteristics of the Arctic mountain grizzly bear. Pages 247–260. *In* M. R. Pelton, J. W. Lentfer, and G. E. Folk, eds. Bears – Their biology and management. Int. Union Conserv. Nat. New Ser. 40. Morges, Switzerland.

12. ROGERS, L. L. 1977.
Social relationships, movements, and population dynamics of black bears in northeastern Minnesota. Ph.D. thesis, University of Minnesota, Minneapolis. 194pp.

13. JONKEL, C. J., AND I. M. COWAN. 1971.
The black bear in the spruce-fir forest. Wildl. Monogr. No. 27. 55pp.

14. BARBER, K. R., AND F. G. LINDZEY. 1986.
Breeding behavior of black bears. Int. Conf. Bear Res. and Manage. 6:129–36.

15. LUDLOW, J. C. 1976.
Observations on the breeding of captive black bears (*Ursus americanus*). Int. Conf. Bear Res. and Manage. 3:65–69.

16. ERICKSON, A. W., AND J. E. NELLOR. 1964.
Breeding biology of the black bear. Mich. State Univ. Agric. Exp. Stn. Res. Bull. 4:5–45.

17. WIMSATT, W. A. 1963.
Delayed implantation in the Ursidae, with particular reference to the black bear (*Ursus americanus* Pallus). Pages 49–76. *In* A. C. Enders, ed. Delayed implantation. The University of Chicago Press, Chicago, Ill.

18. LICHE, H., AND K. WODZICKI. 1939.
Vaginal smears and the oestrous cycle of the cat and lioness. Nature. 144:245–46.

19. HANSSON, A. 1947.
The physiology of reproduction in mink (*Mustela vision* Schreb) with special references to delayed implantation. Acta Zool. 28:1–136.

20. FARRIS, E. J., ED. 1950.
The care and breeding of laboratory animals. John Wiley and Sons, Inc., New York, N.Y. 515pp.

21. ASDELL, S. A. 1964.

Patterns of mammalian reproduction. Cornell University Press, Ithaca, N.Y. 670pp.

22. REYNOLDS, D. G., AND J. J. BEECHAM. 1980.

Home range activities and reproduction of black bears in west central Idaho. Int. Conf. Bear Res. and Manage. 4.

23. POELKER, R. J., AND H. D. HARTWELL. 1973.

Black bear of Washington. State Game Dept. Biol. Bull. 14. 180pp.

24. LINDZEY, F. G. 1976.

Black bear population ecology. Ph.D. thesis, Oregon State University, Corvallis. 105pp.

25. RAUSCH, R. L. 1961.

Notes on the black bear (*Ursus americanus* Passas) in Alaska, with particular reference to dentition and growth. Z.f. Saugetierkunde. 26:65–128.

26. STICKLEY, A. R., JR. 1961.

A black bear tagging study in Virginia. Proc. Ann. Conf. S.E. Game and Fish Comm. 15:43–54.

27. HAMILTON, R. 1972.

Summaries, by state: North Carolina. Pages 11–13. In R. L. Miller, ed. Proceedings of the 1972 black bear conference. N.Y. State Dept. Environ. Conserv. Delmar, N.Y.

28. COLLINS, J. M. 1973.

Some aspects of reproduction and age structures in the black bear in North Carolina. Proc. Ann. Conf. S.E. Assoc. Game and Fish Comm. 27:163–70.

29. RAYBOURNE, J. W. 1976.

A study of black bear populations in Virginia. Trans. Northeast. Sect., The Wildl. Soc., Fish and Wildl. Conf. 33:71–81.

30. KORDEK, W. S., AND J. S. LINDZEY. 1980.

Preliminary analysis of female reproductive tracts from Pennsylvania black bears. Int. Conf. Bear Res. and Manage. 4:159–62.

31. FREE, S. L., AND E. MCCAFREY. 1972.

Reproductive synchrony in the female black bear. Pages 199–206. In S. Herrero, ed. Bears – Their biology and management. IUCN Publ. New Ser. 23.

32. KOHN, B. 1982.
Status and management of black bears in Wisconsin. Tech. Bull. No. 129, Wisconsin Dept. Nat. Res. 31pp.

33. SPENCER, H. E., JR. 1955.
The black bear and its status in Maine. Maine Dept. Inland Fish and Game Bull. 4. 55pp.

34. HARLOW, R. F. 1961.
Characteristics and status of Florida black bear. Trans. North Am. Wildl. Conf. 26:481–95.

35. PIEKIELEK, W., AND T. S. BURTON. 1975.
A black bear population study in northern California. California Fish and Game. 61(1):4–25.

36. ROGERS, L. L. 1976.
Effects of mast and berry crop failures on survival, growth, and reproductive success for black bears. Proc. North Am. Wildl. and Nat. Resour. Conf. 41:431–38.

37. BARNES, V. G., AND O. E. BRAY. 1967.
Population characteristics and activities of black bears in Yellowstone National Park. Final Rep. Colorado Coop. Wildl. Res. Unit, Colorado State University, Fort Collins. 196pp.

38. CAUGHLEY, G. 1977.
Analysis of vertebrate populations. John Wiley and Sons, New York, N.Y. 23pp.

39. AMSTRUP, S. C., AND J. J. BEECHAM. 1976.
Activity patterns of radio-collared black bears in Idaho. J. Wildl. Manage. 40:340–48.

40. ROHLMAN, J. A. 1989.
Black bear ecology near Priest Lake, Idaho. M.S. thesis, University of Idaho, Moscow. 76pp.

41. LECOUNT, A. L. 1982A.
An analysis of the black bear harvest in Arizona (1968–1978). Arizona Game and Fish Dept. Spec. Rep. No. 12. Phoenix. 42pp.
LeCount, A. L. 1982b. Characteristics of a central Arizona black bear population. J. Wildl. Manage. 46:861–68.

42. ERICKSON, A. W., AND G. A. PETRIDES. 1964.
Population structure, movements, and mortality of tagged black

bears in Michigan. Mich. State Univ. Res. Bull. 4:46–67.

43. JOHNSON, K. 1982.
Bait station surveys to determine relative density, distribution, and activities of black bears in the Southern Appalachian Region. Univ. of Tenn. Annu. Prog. Rep.

44. HERRERO, S. 1978.
A comparison of some features of the evolution, ecology, and behavior of the black and grizzly/brown bears. Carnivore. 1:7–17.

TEN | HARVEST CHARACTERISTICS

SHORTLY AFTER MOVING TO IDAHO IN 1968 TO BE-
gin graduate school, I took up black bear hunting. Although I
had always been fascinated by predators, especially bears, I
had never before lived in an area where bears could be hunted.
Back then, the black bear was classified as a game animal, but
hunters didn't need bear tags, and most areas of the state had
year-round seasons. My first bear hunting trips were exciting
and mostly unsuccessful. Finding bear tracks and droppings
along the logging roads outside Elk River wasn't hard –
finding the bear that made them was.

My luck didn't change until spring 1971, when I got to-
gether with two friends from the department for a bear hunt in
southwestern Idaho. The first week of the hunt, Gary Power
and I spent hours scanning open hillsides but couldn't find a
bear to stalk. Discouraged, Gary and I went home for the
weekend to rest and recover. The following week, Gary and I
met Dave Neider, the local biologist, for another few days of
hunting in the same area. To my amazement, we saw 13 bears
in the next three days, and both Gary and I took nice bears.

I've learned much about bear hunting since then, and
now have a better idea why I didn't see any bears one week, but
saw many bears the following week: Even though bears come

out of their winter dens in April, they don't become fully active for two to three weeks (usually during the first week or so of May). I've also learned more about black bear hunting through the department's mandatory check program, which requires successful bear hunters to bring their bears to a department checkpoint. The harvest data gathered from the mandatory check are used to help manage bear populations. To find out what we've learned about bears from these harvest data, please read on.

JOHN BEECHAM

For many years, there were no restrictions on hunting Idaho black bears: Hunters could take as many bears as they liked, using any hunting method, at any time of the year. In 1943, the black bear was classified as a game animal, with a bag limit of one bear during a yearlong season. Few protective regulations were passed until 1973, when resident hunters were required to have a bear tag in their possession while hunting black bears in those game management units (GMUS) in northern Idaho that had summer hunting closures. Resident bear hunters in much of southern Idaho, where seasons were still open to year-round hunting, didn't need a tag. Nonresident black bear hunters were required to have a bear tag in all GMUS in the state.

Since then, the popularity of the black bear as a big game animal has grown considerably. Today, it ranks third behind deer and elk in terms of the number of days of hunting recreation provided to Idaho sportsmen. In 1990, 15,100 hunters spent over 101,000 days in the field hunting black bear. As a result of this growing interest in black bear hunting, we implemented a variety of new regulations to improve our ability to manage this species. These changes included (1) a mandatory check of all bears harvested after 1982, (2) protection for female bears accompanied by young, and (3) summer and winter hunting season closures. We also established harvest criteria and *data analysis units* (DAUS) for monitoring bear harvests.

During our studies, we collected data on hunter numbers, hunting methods used (method of take), types of weapons used, overall harvest trends, and harvest trends by area. We also looked at the effects of hunting pressure on bear populations and examined recommendations for keeping hunting pressure within reasonable limits.

DATA COLLECTION

To collect harvest information, we relied on two primary methods: the mandatory check and report program, and the annual telephone harvest survey. The mandatory check and report program, implemented in 1983, required each successful hunter to bring the skull of his/her black bear to an official checkpoint within ten days of killing the bear and to fill out a report form. In most cases, we extracted a tooth from the skull for aging purposes. Pertinent data including the kill date, location of kill, method of take, and sex of the bear were recorded.

The annual telephone harvest survey, started in 1979, provided a second estimate of the black bear harvest. The telephone survey crew contacted about 3 percent of the bear tag holders and collected information from both successful and unsuccessful black bear hunters. We were also able to use the phone survey to estimate hunter success and the number of recreation days sportsmen spent hunting for black bears each year.

From 46 to 69 percent (average 59 percent) of Idaho hunters complied with the mandatory check and report regulation during 1983 to 1990. As a result, the number of bears checked annually represents the minimum number of bears harvested that year. Compliance with the reporting regulation was probably poorest among incidental hunters (sportsmen who shot bears while hunting for other animals or while cutting firewood, picnicking, or berry picking). Often they weren't as familiar with the black bear mandatory check as hunters who were specifically hunting for bears. During 1989 and 1990, compliance with the mandatory check improved, probably because of a decline in the number of incidental hunters in recent years and an increase in hunters targeting black bears as their primary quarry.

HUNTER NUMBERS

From 1983 to 1990, data collected from our annual telephone harvest survey and department black bear tag sales showed differing
trends in the number of black bear hunters. The telephone survey
showed a gradual increase in bear hunters (Fig. 10–1). However, ac-

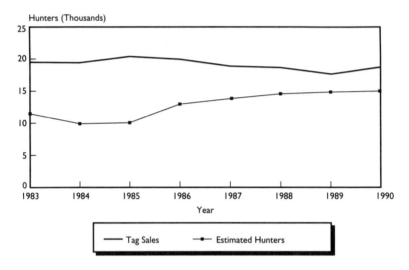

Figure 10–1. Black bear tag sales and estimated number of black bear hunters
(from telephone survey), 1983–90.

tual black bear hunting tag sales decreased during this same time
frame. Sales of resident black bear tags remained stable, but nonresident tag sales dropped dramatically from 4,509 in 1986 to 2,708
in 1989 (Table 10–1). The sale of nonresident bear tags fell notice-
ably when tag prices were increased from $25.50 to $40.50 in 1987.
Another significant drop occurred in 1989 when nonresident tag
prices went up to $91.00. Despite these dramatic drops in nonresi-
dent tag sales, the number of nonresident bear hunters who used
outfitter and guide services increased 66 percent from 1986 to 1990.
    Although the tag sale numbers and telephone survey data were
contradictory, we believe the telephone harvest survey data indicate that interest in black bear hunting is increasing in Idaho. The

Table 10–1. History of black bear tag sales, 1973–90

Year	Resident Bear Tag Sales	Cost of Resident Bear Tag	Resident Extra Tag Sales[1]	Cost of Extra Bear Tag	Nonres. Bear Tag Sales	Cost of Nonres. Bear Tag	Nonres. Extra Tag Sales[1]	Cost of Nonres. Extra Tag
1973	14,445	2.00			758[2]	25.00[2]		
1974	17,302	2.00			829[2]	25.00[2]		
1975	10,672	4.00	66[3]	4.00	2,205	15.00	15[3]	15.00
1976	13,453	4.00	408	4.00	2,903	15.00	133	15.00
1977	13,328	4.00	429[4]	4.00	3,207	15.00	152[4]	15.00
1978	13,738	4.00	374[5]	4.00	4,032	15.00	219[5]	15.00
1979	15,233	4.50	374[5]	4.50	4,738	15.50	253[5]	15.50
1980	15,972	4.50	393	4.50	4,773	15.50	255	15.50
1981	15,944	5.50	466	5.50	4,754	20.50	205	20.50
1982	14,682	6.50	236	6.50	3,873	25.50	121	25.50
1983	15,252	6.50	275	6.50	4,341	25.50	112	25.50
1984	15,033	6.50	278	6.50	4,476	25.50	155	25.50
1985	15,456	6.50	261	6.50	4,978	25.50	154	25.50
1986	15,464	6.50	193	6.50	4,509	25.50	110	25.50
1987	14,954	6.50	–	–	3,928	40.50	–	–
1988	15,120	6.50	–	–	3,601	40.50	–	–
				(Effective 7/1/88)		90.50		
1989	15,034[6]	7.00	–	–	2,708	91.00	–	–
1990	15,934	7.00	–	–	2,848	91.00	–	–

[1] This policy started in 1975 as part of a calf elk mortality study.

[2] Nonresidents bought nonresident bear tags until 1975 when they were required to buy a nonresident game license and bear tag.

[3] Used to take one black bear in units 10, 12, or 16.

[4] Used to take one black bear in units 8, 8A, 9A, 10, 10A, 12, 15, 16, 16A, 17, 19, 20, 20A, 23, 24, 25, 26, 34, and 39.

[5] Used to take one black bear in units 8, 8A, 9A, 10, 10A, 12, 15, 16, 16A, 17, 19, 20, 20A, 25, 26, 34, and 39 (two bears could be taken without tags in units 19A, 23, 24, 33, 35, 43, and 44).

[6] This number does not include bear tags sold as part of the Sportsman Package.

decrease in tag sales probably reflected a drop in the number of deer and elk hunters who were buying bear tags on the chance that they might see a bear while hunting in the fall. If this is true, the actual increase in sportsmen who are specifically hunting for black bears may be more significant than estimated from our telephone harvest survey. Tags sold in the *Sportsman Package* in 1989 (2,185) and 1990 (2,968) weren't included in these tag sales data because we didn't know how many of those hunters intended to use their bear tags and because few Sportsman Packages were sold.

METHOD OF TAKE

Hunters used a variety of techniques to pursue black bears in Idaho. For record-keeping purposes, we put these techniques into four categories: (1) still hunting – using binoculars to locate bears feeding in open areas and then stalking them; (2) hound hunting – using trailing dogs to pursue and "tree" bears; (3) bait hunting – using food or scents to attract bears to a specific spot; and (4) incidental hunting – coming across black bears while engaged in some other recreational activity, such as hunting for other game species, fishing, wood cutting or berry picking. Occasionally, hunters used a combination of techniques (for example, houndsmen often "strike" bears off baits¹), but they reported only one method. We couldn't evaluate how extensive these practices were, so all our analyses reflected only those data reported on each mandatory report form.

The proportion of hunters using different methods of take remained relatively constant between 1983 and 1990. Still hunters killed 32 percent of the bears harvested during the eight-year period; incidental hunters, 29 percent; hound hunters, 20 percent; and bait hunters, 19 percent.

Even though bait hunters took the smallest proportion of the bears harvested, we documented that bait hunters were becoming more successful. From 1983 to 1990, bait hunters showed the largest increase in harvest (118 percent), followed by hound hunters (97 percent) and still hunters (64 percent). Although hunters using bait and hounds comprised the smallest group of black bear hunters (11 and 5 percent, respectively, in 1990), they consistently had the highest success rates (40 to 50 percent). Since 1986, bait hunters have surpassed hound hunters: In 1990, hunters using bait harvested more black bears than hound hunters (Fig. 10–2). In addition, the proportion of bait hunters almost doubled between 1984 and 1990 (Fig. 10–3), while the percentage of hound hunters stayed essentially the same (1).

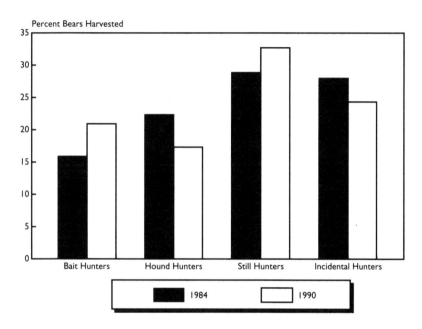

Figure 10–2. Proportion of bears harvested by bait, hound, still, and incidental hunters, 1984 and 1990.

An analysis of *hunter effort* records from our mandatory report indicated that hunters using bait took longer to kill a bear than did other types of hunters (Table 10–2). However, between 1983 and 1990, hunter effort decreased for bait hunters, increased for still and hound hunters, and stayed roughly the same for incidental hunters.

WEAPON TYPES USED

Black bear hunters in Idaho used several types of weapons for pursuing bears: rifles, bows and arrows, muzzleloaders, and pistols.

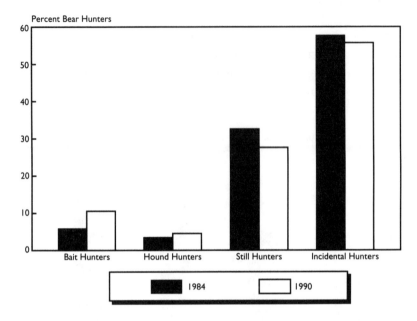

Figure 10–3. Proportion of bait, hound, still, and incidental hunters, 1984 and 1990.

Table 10–2. Trends in the effort required to harvest a black bear in Idaho by method of take, 1983–90

	Number of Days/Bear Harvested			
Year	Still	Incidental	Bait	Hounds
1983	3.1	2.4	6.3	2.7
1984	3.3	2.9	6.0	2.9
1985	4.1	2.7	5.5	2.9
1986	3.5	2.9	6.1	2.8
1987	3.5	3.5	6.1	3.0
1988	3.4	3.3	6.3	3.5
1989	3.6	2.7	4.9	4.8
1990	3.7	2.8	5.3	4.2
Average	3.5	2.9	5.8	3.4

From 1983 to 1990, rifle hunters killed the largest number of black bears (Table 10–3), taking an average of 3.4 days (range 3.1 to 4.0 days) to get one bear. The number of bears taken by rifle hunters increased 56 percent between 1983 and 1990.

Archery hunters took the second highest number of bears. To get their quarry, they hunted an average of 5.0 days (range 3.3 to 6.3 days) using still hunting methods and 6.8 days (range 6.0 to 7.4 days) using bait. The number of bears taken by archers increased 87 percent between 1983 and 1990.

Hunters using muzzleloaders and pistols accounted for a small number of the black bears harvested during 1983–1990. However, pistol hunters took 122 percent more bears in 1990 than they did in 1983.

The number of bears harvested by rifle hunters, archers and pistol hunters increased significantly during 1983–1990 (Table 10–3). Because of small sample sizes, we couldn't tell if harvest numbers increased or decreased for muzzleloader hunters. No trends were apparent in the number of days it took to harvest a bear using different weapon types. However, our data indicated that it took less time to kill a bear with a rifle than with other methods, and archers who used still hunting methods took significantly less time to take a bear than archers using bait. We observed a significant rise in the number of bears taken by rifle hunters using bait.

Analysis of our data showed no trend in the average age or median age of black bears taken by various weapon types or hunting methods. Hound hunters harvested slightly older bears than hunters using other methods (Table 10–4).

EFFECTS OF HUNTING PRESSURE

The intensity of hunting pressure influences bear harvest rates, population age structures, and ultimately, the number of bears. Population models that simulate different harvest strategies suggest that as harvest rates increase, we can expect more females to be harvested. The average age of harvested males will also decline, while that of females will rise slightly (2, 3). Harris (2) reported that heavily hunted populations first showed increases in the proportion

Table 10–3. Number of black bears checked in Idaho by weapon type, 1983–90

| | Weapon Type | | | |
Year	Rifle	Bow	Pistol	Muzzleloader
1983	781	126	44	3
1984	806	144	68	9
1985	874	144	61	3
1986	961	194	74	7
1987	722	168	68	4
1988	880	183	76	8
1989	1,091	201	95	12
1990	1,218	235	98	8
Average	917	174	73	7

Table 10–4. Number of black bears checked and average age by method of take and weapon type in Idaho, 1983–90

| | Still Hunt | | | | Bait[1] | | | | Hounds | | | |
| | Rifle | | Bow | | Rifle | | Bow | | Rifle | | Bow | |
Year	Bears	Age	Bears	Age	Bears	Age	Bears	Age	Bears	Age	Bears	Age
1983	227	4.4	7	4.7	51	4.4	78	4.2	117	4.9	12	4.8
1984	197	4.4	15	3.7	61	4.6	65	3.9	117	4.4	21	5.2
1985	244	3.5	15	3.7	72	4.5	64	3.7	115	4.2	22	5.5
1986	274	4.1	15	3.1	66	4.0	96	4.4	59	4.7	21	4.9
1987	172	4.4	15	4.7	65	4.0	86	4.7	69	4.8	13	5.0
1988	218	4.6	12	4.7	85	4.5	90	4.4	103	5.7	36	4.6
1989	294	4.3	15	4.8	134	4.1	87	3.7	116	5.0	37	5.5
1990	379	5.0	28	4.4	148	5.2	121	4.9	158	5.5	32	5.1

[1] Significant upward trend in numbers between 1983 and 1990.

Interest in black bear hunting seems to be increasing in Idaho. Photo by Jeff Rohlman

of adult males. If heavy hunting pressure continued, the proportions of subadult males and females rose, as older males were depleted from the population. At even higher harvest levels, the number of females taken increased.

The population may then decline in total numbers unless there are "reservoir" areas nearby that can produce subadult bears to repopulate these heavily hunted areas (4). In these areas, populations are regulated by the capacity of the habitat to support bears and behavioral factors. Young male black bears (but not usually young females) produced in reservoir areas leave their mother's home range at 1.5 to 2.5 years of age and may travel long distances to repopulate vacant bear habitat. As a result, in heavily hunted populations located near reservoir areas, hunters often take high numbers of these traveling subadult males. In heavily hunted populations that aren't near reservoir areas, hunters end up taking increasing numbers of older females. Bear populations face another problem in isolated areas that are heavily hunted: Because females usually don't disperse far from their mother's home range, the females lost in these heavily hunted populations probably won't be replaced by dispersing females from reservoir areas. These trends were supported by our research studies in Idaho (see chapter 9) as well as studies conducted in Alaska (5), Virginia (6), and Arizona (7).

The vulnerability of black bears to hunting varies greatly within Idaho because of differences in habitat and hunter access to those habitats. Bears are generally less vulnerable to harvest where cover is dense over large areas. They are more vulnerable in open or patchy habitats. The accessibility of an area also influences how many bears hunters take: Black bear populations in areas with many roads are more prone to heavy harvests than are populations in unroaded or wilderness areas. As a result, accessible areas often contain fewer adult males than reservoir areas.

The sex and age of black bears also affects their vulnerability to hunters. Adult males (those four years or older) are the most vulnerable because of their bold, aggressive behavior and large home range sizes (about 30 to 35 square miles). Subadult males are somewhat less vulnerable than adult males, but are more vulnerable than females. Adult females, especially those accompanied by cubs, are the least vulnerable to hunters (8, 7, 3). When method of take is considered, this pattern changes somewhat. Female black bears were more vulnerable to hunters using hounds in Washington (9) and Arizona (7) than to hunters using other hunting methods.

The vulnerability of bears to hunting influences the age structure of a population. Caughley (10) demonstrated that age structures were unaffected by increasing or decreasing mortality rates when these rates were applied equally to all ages. He also found that, in theory, age structures were similar for increasing and decreasing populations if the animals were all equally vulnerable to hunters. However, this situation doesn't occur in real life; Harris (2) showed that bears had different vulnerabilities to hunters. He also noted that *harvest age structures* depended on the age structures of bear populations and on how vulnerable bears of different sex and age classes were to hunters. Harris demonstrated that if harvest remained constant as the vulnerable age classes were depleted, the less vulnerable bears (adult females) would be taken. He reported that harvest age structures were highly variable for black bears and responded very slowly to changing harvest rates. Miller and Miller (3) suggested that managers should be especially cautious about assuming that a bear population is stable because the average of age of its members hasn't changed.

The number of two-year-old bears harvested and checked by still and incidental hunters, however, may give us an idea of bear population productivity and stability. This number varied considerably from year to year. If this harvest number is strongly correlated with the number of cubs produced two years previously, it may be worthwhile to use it as an index to the productivity of a population.

We've looked at how hunting pressure affects bear populations. Now consider another question: How fast will reductions in hunting pressure affect bear populations? Harris (2) and Miller and Miller (3) reported that because of their low reproductive rates, populations were slow to respond to hunting regulation changes designed to reduce harvest. They also indicated that small adjustments in harvest rates meant long recovery periods for populations, even assuming high reproductive rates and low mortality rates. Miller's (11) modeling showed that it would take black bear populations 17 years to recover from a 50 percent population loss if hunting pressure was reduced by 25 percent.

OVERALL HARVEST TRENDS

In analyzing our harvest data for 1983–1990, we noted the following trends: The number of black bears hunters reported killing increased significantly – 66 percent (Fig. 10–4). When we pooled all data by method of take, we also saw it was taking longer for hunters to kill a bear. In addition, we noticed that hunters were taking more females, especially older ones. These increasing harvest trends persisted despite hunting season changes in 1986 that were designed to reduce harvest. This suggested that on a statewide basis we should monitor future harvests closely for increasing signs of overharvest.

However, we couldn't detect any change in the proportion of adult males or females harvested or in the number of bears (males or females) taken that were older than ten years. The median age of

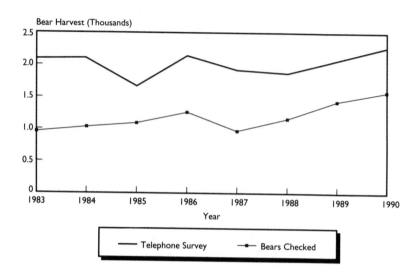

Figure 10–4. Black bear harvest as estimated by telephone survey and mandatory check, 1983–90.

harvested males (three) and females (four) hasn't changed. The mean age of harvested males also stayed the same, but that of harvested females rose (Table 10–5). When we grouped harvested bears by years, 1983–1986 and 1987–1990, we saw no significant differences in the age distribution for females. Male bears checked during 1987–1990, however, were older than males checked in 1983–1986.

We interpreted these conflicting trends in the harvest data to mean that harvest was occurring within acceptable limits on a statewide basis. Yet some data suggested that the situation was changing, especially during the past couple years. Interest in black bear hunting was rising, harvest was increasing, and it was taking longer to kill a black bear. Harris (2) stated that the symptoms of overharvest weren't always obvious until a population decline was well under way. He advised managers to make decisions about managing bears based on the risk of overharvest rather than waiting for solid, scientific evidence of overharvest. Although we believe we were starting to see some symptoms of overharvest, we weren't able

Table 10–5. Age structure and sex composition of black bears checked in harvest, 1983–90

Year	Number of Males and Females M	F	Ratio of Males to Females	Age Median M	F	Mean M	F	Proportion of Harvest (%) Adult Male	Adult Female
1983	639	318	2.0	3	4	4.4	4.6	34	15
1984	657	373	1.8	3	4	3.9	4.6	27	19
1985	668	420	1.6	2	3	3.4	4.3	23	17
1986	811	431	1.9	3	4	3.9	4.8	27	17
1987	636	335	1.9	3	4	4.1	5.0	32	19
1988	744	404	1.8	3	4	4.3	5.2	32	19
1989	964	466	2.1	3	4	3.8	4.5	29	17
1990	1,003	576	1.7	3	4	4.6	5.4	32	20

to document a significant population decline between 1983 and 1990.

Our harvest data showed some other interesting results. Black bear hunters took 53 percent of the harvested bears during the spring season. From 1983 to 1990, harvests increased 95 percent during the spring, but only 47 percent during the fall (Tab. 10–6). These data can probably be explained by a two-week hunting season closure in late September 1986–1990, throughout much of the state. We were surprised to see a significant increase in the number of female black bears harvested during the spring (110 percent) and fall seasons (59 percent). Because Idaho experienced droughtlike conditions during four of the past five years, we expected all bears to be more vulnerable to hunters, especially during fall. These data suggested that although some of the increase in fall harvest was probably due to drought, the large increase in spring harvest was due to increased hunting pressure.

The harvest data also showed that more males than females were taken by hunters. We divided the spring and fall seasons into ten two-week periods and examined the sex ratio of the harvest.

Table 10–6. Distribution of checked bears, by sex, during the spring and fall seasons, 1983–90

	Spring			Fall		
Year	*Males*	*Females*	*Total*	*Males*	*Females*	*Total*
1983	320	131	451	319	187	506
1984	322	195	517	333	178	511
1985	314	170	484	354	219	573
1986	426	194	620	377	232	609
1987	319	153	472	316	179	495
1988	393	173	566	345	228	573
1989	506	215	721	448	246	694
1990	556	275	831	439	297	736

When the data were pooled, males outnumbered females more than 2:1 in the spring and 1.5:1 in the fall. These ratios remained relatively constant from one two-week period to the next in both spring and fall. In addition, our analysis of the absolute numbers of bears taken showed that large numbers of females were harvested in the last two weeks of May and the first two weeks of September. A third increase in female harvest occurred in early October, coinciding with the opening of the general rifle season for elk (Fig. 10–5). The distribution of the male harvest over these two-week intervals was the same as that of the females.

HARVEST TRENDS BY AREA

Idaho is divided into five areas for managing black bear populations (Fig. 10–6). Two of the five areas (areas 1 and 4) were broken down into smaller groups so we could better analyze harvest information. These small groups had similar habitats, road access, and proximity to urban population centers, and were called data analysis units (DAUS). To determine black bear harvest trends in bear management areas, we looked at the percent of females in the harvest and the median age of harvested males and females. These harvest criteria (Table 10–7) are used by the Idaho Department of Fish and Game to monitor black bear harvest and are outlined in the department's

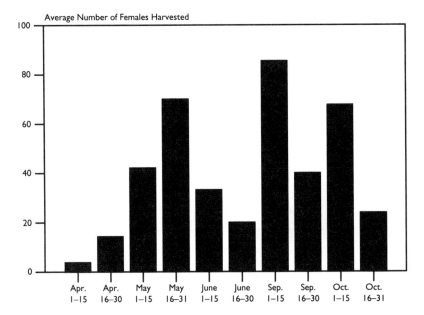

Figure 10–5. Distribution of the female black bear harvest, 1983–90 (by two-week intervals).

1992–2000 Black Bear Management Plan. We saw the following harvest trends in each area:

Area 1   This area contained ten DAUS and consisted of black bear habitat that varied from dense, semicoastal forests in the Idaho panhandle to dry river breaks with patchy timber stands in southern Idaho. Because area 1 had many roads, most black bear habitat in these DAUS was readily accessible to the public.

Our analysis of the sex composition and age structure information we collected from checked bears during 1983 through 1990 showed that 68 percent of the black bear harvest came from area 1. No DAUS in area 1 met the desired female median age (six years or older), and only three DAUS met the desired male median age of four years or older (Table 10–8). Four of ten DAUS met the desired level of 35 percent or less females in the harvest. Data on hunting effort, sex composition, and age structure for each DAU by year are presented in Tables 10–9 through 10–18.

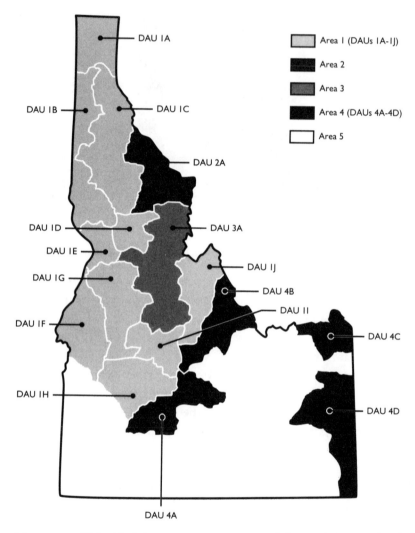

Figure 10–6. Idaho black bear management areas and data analysis areas (DAUS). See Table 10–8 for game management units in each DAU.

Table 10–7. Criteria used to monitor black bear harvest in Idaho[c]

Criteria	Overharvest[a]	Desired Level[b]
Percent females	≥40	≤35
Median age	≤3 years	≥5 years
Males	≤2 years	≥4 years
Females	≤4 years	≥6 years
Bait station survey	Declining	Stable or increasing

[a] Reflects an overharvested population.
[b] Relects a self-sustaining, viable population with a diverse age structure.
[c] Taken from the Idaho Department of Fish and Game 1992–2000 Black Bear Management Plan.

Area 2   Area 2 was made up of habitats similar to those found in the northern part of area 1. This area was less accessible by road than area 1 and wasn't close to human population centers. Area 2 accounted for 16 percent of the state's black bear harvest. Most of the harvest, except that taken by deer and elk hunters, probably occurred along road or river corridors. As a result, much of this area probably functioned as a reservoir area, producing many of the subadult males that moved into vacant habitats in adjacent DAUs.

Area 2 didn't meet the desired median ages for males or females during 1983–1990, but did have less than 35 percent females in the harvest (Table 10–8). We weren't concerned about these data, because area 2 contained large reservoir areas and we believed much of the harvest was focused on young, dispersing animals rather than on the core population that resided in relatively inaccessible habitats away from roads. Our harvest data came from the more heavily hunted road or river corridors, and probably didn't represent the area as a whole. Data on hunting effort, sex composition, and age structure for DAU 2A are presented in Table 10–19.

Area 3   This area contained habitats similar to those found in areas 1 and 2, but access was very limited and human populations were low. Much of area 3 lay in remote habitats in two central Idaho wilderness areas. Although hunting pressure was low, in recent years it

Table 10–8. Harvest characteristics for checked black bears by DAU, 1983–90

DAU	Units	1983–90 Harvest	Percent Female	Median Age Male	Median Age Female
Area 1					
1A	1	961	37	3	4
1B	2, 3, 5, 8, 11, 11A	847	38	3	3
1C	4, 4A, 6, 8A, 10A	1,522	36	3	3
1D	15, 16	239	34	4	4
1E	13, 14, 18	273	27	3	4
1F	22, 31, 32, 32A	393	42	2	4
1G	19A, 23, 24, 25, 33	984	33	3	4
1H	39, 43	573	39	3	3
1I	34, 35, 36	162	24	4	5
1J	21, 21A, 28, 36B	478	37	4	4
Area 2					
2A	7, 9, 10, 12	1,530	32	3	5
Area 3					
3A	16A, 17, 19, 20, 20A, 26, 27	588	35	4	4
Area 4					
4A	44, 45, 48, 49	143	34	3	3
4B	29, 30, 30A, 36A, 37, 37A	347	33	4	4
4C	60, 61, 62, 62A	125	29	4	4
4D	64, 65, 66, 66A, 67, 69, 76	262	44	3	3

has increased along the two major river corridors (the Salmon and Selway rivers).

About 6 percent of the black bear harvest occurred in area 3. The 1983–1990 median age of both male and female checked bears was four years and the average percent females in the harvest was 35 (Table 10–8). Although these data indicated that black bear populations were hunted at moderately heavy rates, we believed that the harvest was dominated by young bears living along river corridors – not bears in the core population. Therefore, our harvest data probably didn't represent the whole area. Many area 3 habitats

Table 10–9. Hunting effort, sex composition, and age structure of checked black bears harvested in DAU 1A, 1983–90

Year	Number of Days/Bear Harvested	Males			Females		
		Number of Bears Aged	Mean Age	Median Age	Number of Bears Aged	Mean Age	Median Age
1983	2.7	69	4.5	3	9	4.8	3
1984	2.8	53	4.8	4	27	5.1	5
1985	3.5	43	2.9	3	31	3.9	3
1986	2.7	65	4.3	4	40	4.9	4
1987	3.8	32	4.4	3	11	5.5	4
1988	3.5	34	4.5	4	11	6.0	5
1989	2.8	73	4.3	3	21	5.8	4
1990	3.7	61	4.6	3	39	5.2	4

Table 10–10. Hunting effort, sex composition, and age structure of checked black bears harvested in DAU 1B, 1983–90

Year	Number of Days/Bear Harvested	Males			Females		
		Number of Bears Aged	Mean Age	Median Age	Number of Bears Aged	Mean Age	Median Age
1983	3.8	35	3.4	3	21	3.6	2
1984	4.2	28	4.0	3	13	4.1	4
1985	3.1	48	2.9	2	21	3.1	2
1986	3.7	54	3.3	2	33	4.8	5
1987	3.4	49	3.3	3	23	5.4	6
1988	4.2	61	3.7	3	35	4.6	3
1989	4.0	66	3.7	3	40	3.6	3
1990	3.5	79	3.4	2	44	5.8	4

Table 10–11. Hunting effort, sex composition, and age structure of checked black bears harvested in DAU 1C, 1983–90

Year	Number of Days/Bear Harvested	Males			Females		
		Number of Bears Aged	Mean Age	Median Age	Number of Bears Aged	Mean Age	Median Age
1983	2.6	85	4.5	3	37	5.4	5
1984	2.9	76	3.5	3	43	4.9	4
1985	3.5	91	2.8	2	53	4.3	2
1986	3.4	102	3.6	3	64	4.6	3
1987	4.6	68	4.3	3	35	4.0	2
1988	4.4	85	4.2	3	47	4.8	3
1989	3.5	118	3.5	2	49	4.1	4
1990	4.1	134	4.1	3	76	5.3	4

Table 10–12. Hunting effort, sex composition, and age structure of checked black bears harvested in DAU 1D, 1983–90

Year	Number of Days/Bear Harvested	Males			Females		
		Number of Bears Aged	Mean Age	Median Age	Number of Bears Aged	Mean Age	Median Age
1983	2.5	13	4.8	3	2	4.0	4
1984	4.0	12	4.5	4	7	4.3	5
1985	5.5	3	5.0	2	8	3.5	3
1986	3.2	12	4.0	3	3	2.7	2
1987	3.6	13	3.2	3	4	6.0	6
1988	3.3	10	3.8	3	97	6.7	8
1989	3.8	18	5.2	4	10	3.9	4
1990	3.6	28	5.1	4	15	5.4	3

Table 10–13. Hunting effort, sex composition, and age structure of checked black bears harvested in DAU 1E, 1983–90

Year	Number of Days/Bear Harvested	Males			Females		
		Number of Bears Aged	Mean Age	Median Age	Number of Bears Aged	Mean Age	Median Age
1983	2.8	13	4.3	4	5	5.0	4
1984	4.1	18	3.3	2	12	5.2	4
1985	2.4	34	3.2	2	7	5.0	4
1986	4.2	21	3.9	3	2	7.0	7
1987	3.3	28	3.6	3	6	3.7	3
1988	5.8	11	5.1	5	5	6.6	6
1989	3.6	22	3.5	3	12	4.8	3
1990	4.2	16	4.8	4	4	4.0	4

Table 10–14. Hunting effort, sex composition, and age structure of checked black bears harvested in DAU 1F, 1983–90

Year	Number of Days/Bear Harvested	Males			Females		
		Number of Bears Aged	Mean Age	Median Age	Number of Bears Aged	Mean Age	Median Age
1983	2.9	15	5.1	4	6	3.8	3
1984	3.5	14	2.1	2	10	3.7	4
1985	3.7	20	3.2	2	17	4.4	4
1986	3.3	11	3.4	3	2	4.0	4
1987	2.6	22	4.2	2	15	4.5	3
1988	2.8	26	4.8	4	25	5.2	4
1989	3.4	32	3.8	2	22	5.6	4
1990	2.5	27	3.9	2	17	3.7	4

Table 10–15. Hunting effort, sex composition, and age structure of checked black bears harvested in DAU 1G, 1983–90

Year	Number of Days/Bear Harvested	Males			Females		
		Number of Bears Aged	Mean Age	Median Age	Number of Bears Aged	Mean Age	Median Age
1983	4.6	44	4.0	4	18	4.9	3
1984	3.3	69	4.1	3	40	3.8	3
1985	2.9	75	3.4	2	40	5.2	4
1986	4.7	42	4.3	3	15	5.1	4
1987	4.1	56	4.4	4	19	5.6	4
1988	4.7	68	4.4	4	24	5.8	4
1989	3.3	81	3.7	3	29	3.9	4
1990	3.8	87	4.7	3	43	5.5	5

Table 10–16. Hunting effort, sex composition, and age structure of checked black bears harvested in DAU 1H, 1983–90

Year	Number of Days/Bear Harvested	Males			Females		
		Number of Bears Aged	Mean Age	Median Age	Number of Bears Aged	Mean Age	Median Age
1983	2.9	24	3.5	2	8	4.0	4
1984	3.2	22	3.4	2	13	3.3	3
1985	3.1	22	3.9	3	22	4.2	4
1986	2.9	26	2.9	2	16	4.4	3
1987	3.0	15	2.5	2	16	4.6	4
1988	2.9	29	3.8	3	22	4.3	4
1989	3.0	69	3.5	2	31	4.2	4
1990	4.3	65	4.8	4	40	3.7	3

Table 10–17. Hunting effort, sex composition, and age structure of checked black bears harvested in DAU 1I, 1983–90

Year	Number of Days/Bear Harvested	Males			Females		
		Number of Bears Aged	Mean Age	Median Age	Number of Bears Aged	Mean Age	Median Age
1983	4.6	9	4.9	4	1	–	–
1984	2.8	12	3.4	3	2	5.0	5
1985	2.4	4	6.3	6	0	–	–
1986	3.5	11	2.9	3	1	–	–
1987	4.8	13	5.6	4	2	4.0	4
1988	2.8	15	5.3	6	4	7.3	7
1989	3.0	9	5.8	6	7	5.3	5
1990	4.8	23	7.3	5	6	8.8	9

Table 10–18. Hunting effort, sex composition, and age structure of checked black bears harvested in DAU 1J, 1983–90

Year	Number of Days/Bear Harvested	Males			Females		
		Number of Bears Aged	Mean Age	Median Age	Number of Bears Aged	Mean Age	Median Age
1983	3.1	27	4.3	4	10	4.6	3
1984	3.0	20	3.3	3	16	6.0	3
1985	3.4	31	5.3	5	9	5.0	4
1986	3.3	34	5.1	5	18	4.6	3
1987	2.8	11	5.5	4	8	5.4	4
1988	3.1	43	4.6	3	20	4.5	4
1989	2.8	23	4.2	4	10	4.9	5
1990	2.9	33	5.2	4	26	6.9	6

During 1983–1990, it took rifle hunters an average of 3.4 days to get a bear.
Photo by Jeff Rohlman

served as reservoir areas for adjacent DAUs. Because of the low number of people in the area and limited access, we believed black bear populations weren't threatened by overharvest. Data on hunting effort, sex composition, and age structure for DAU 3A are presented in Table 10–20.

Area 4   Area 4 contained four DAUs in southern Idaho. These black bear habitats were marginally suitable for bears because they occurred in small patches and had few berry-producing shrubs. The domestic sheep industry is a major user of public lands in area 4; sheepherders frequently shot black bears to protect their flocks from bear depredations.

Area 4 accounted for about 10 percent of the black bear harvest in Idaho. Although none of the DAUs in area 4 met the desired female median age, and only two met the desired male median age, only DAU 4D had more than 35 percent females in the harvest

Table 10–19. Hunting effort, sex composition, and age structure of checked black bears harvested in DAU 2, 1983–90

Year	Number of Days/Bear Harvested	Males			Females		
		Number of Bears Aged	Mean Age	Median Age	Number of Bears Aged	Mean Age	Median Age
1983	3.9	87	4.8	4	27	5.4	5
1984	3.9	71	4.3	4	36	4.6	4
1985	3.3	82	3.4	2	35	4.3	3
1986	3.9	105	4.0	3	38	4.4	4
1987	4.0	73	4.4	4	33	6.1	6
1988	4.3	102	4.4	3	52	5.7	4
1989	4.1	135	3.8	3	58	5.1	5
1990	3.6	147	4.9	4	61	6.0	5

Table 10–20. Hunting effort, sex composition, and age structure of checked black bears harvested in DAU 3, 1983–90

Year	Number of Days/Bear Harvested	Males			Females		
		Number of Bears Aged	Mean Age	Median Age	Number of Bears Aged	Mean Age	Median Age
1983	3.9	23	4.6	4	11	4.5	4
1984	4.5	34	4.6	5	19	5.2	4
1985	4.0	44	4.4	2	18	5.4	4
1986	3.8	39	3.8	3	23	5.3	4
1987	3.6	43	4.1	3	19	6.3	7
1988	3.5	44	4.4	4	20	6.1	5
1989	3.4	53	4.6	4	25	4.9	4
1990	4.6	57	5.5	4	36	5.8	5

Table 10–21. Hunting effort, sex composition, and age structure of checked black bears harvested in DAU 4A, 1983–90

Year	Number of Days/Bear Harvested	Males			Females		
		Number of Bears Aged	Mean Age	Median Age	Number of Bears Aged	Mean Age	Median Age
1983	3.3	6	4.2	2	1	–	–
1984	2.3	6	2.5	2	4	6.3	4
1985	2.6	7	3.1	2	4	4.0	3
1986	4.7	11	3.4	2	5	5.4	7
1987	6.5	13	3.7	3	2	3.5	3
1988	3.2	9	3.8	3	4	4.0	2
1989	4.4	12	4.5	3	8	2.9	3
1990	5.2	17	3.2	2	16	5.4	5

Table 10–22. Hunting effort, sex composition, and age structure of checked black bears harvested in DAU 4B, 1983–90

Year	Number of Days/Bear Harvested	Males			Females		
		Number of Bears Aged	Mean Age	Median Age	Number of Bears Aged	Mean Age	Median Age
1983	3.6	20	4.7	4	5	3.0	3
1984	4.3	20	3.1	3	13	4.5	4
1985	6.3	23	4.5	4	15	3.9	3
1986	3.5	28	4.5	3	3	5.0	2
1987	3.6	13	4.1	3	13	4.1	3
1988	4.8	33	4.8	4	14	5.5	5
1989	5.0	24	4.3	4	7	6.1	4
1990	3.8	28	4.6	4	9	5.8	5

Table 10–23. Hunting effort, sex composition, and age structure of checked black bears harvested in DAU 4C, 1983–90

Year	Number of Days/Bear Harvested	Males			Females		
		Number of Bears Aged	Mean Age	Median Age	Number of Bears Aged	Mean Age	Median Age
1983	5.0	12	5.3	3	7	2.6	2
1984	6.4	13	4.2	4	1	–	–
1985	10.4	5	4.0	3	2	7.5	7
1986	3.8	14	3.9	3	1	–	–
1987	4.0	2	2.0	2	3	5.3	5
1988	3.4	12	4.6	4	6	4.0	3
1989	5.8	6	2.3	2	3	6.7	7
1990	4.6	8	4.4	3	1	–	–

Table 10–24. Hunting effort, sex composition, and age structure of checked black bears harvested in DAU 4D, 1983–90

Year	Number of Days/Bear Harvested	Males			Females		
		Number of Bears Aged	Mean Age	Median Age	Number of Bears Aged	Mean Age	Median Age
1983	4.9	17	2.5	3	7	4.1	4
1984	7.1	14	3.1	3	10	4.0	3
1985	7.0	10	2.4	2	12	5.5	5
1986	9.5	17	4.2	3	11	5.2	4
1987	8.1	13	4.8	5	10	3.5	2
1988	6.1	11	2.5	3	9	4.2	3
1989	6.9	11	3.5	3	16	3.0	2
1990	6.7	25	3.8	3	9	3.0	3

(Table 10–8). Black bear harvests in area 4 were low and harvest data were variable. As a result, we were reluctant to draw any conclusions from the data. Harvest data on hunter effort, sex composition, and age structure are presented in Tables 10–21 through 10–24.

Area 5   Area 5 contained most of the irrigated lands in southern Idaho and the drier, desertlike habitats. Because habitat quality for black bears was marginal, few black bears were found in area 5. They didn't occur in sufficient numbers to justify an open hunting season.

MANAGEMENT RECOMMENDATIONS

As a result of our research and management actions over the past 20 years, a great deal of information is available about black bear populations and their response to harvest. However, we still have much to learn about this secretive and resourceful species. The modeling studies conducted by Harris (2) and Miller and Miller (3) were very useful in identifying how bear populations responded to various harvest levels. These studies showed patterns that were consistent with other data found by capture-mark-recapture studies. When bear populations experience high harvest rates, adult males become scarce, and more females and younger males are harvested. In addition to these patterns, we also expect to see hunter success rates decline or hunting effort increase. The tendency for young male black bears to disperse as two-year-olds into vacant habitat and the availability of large, remote reservoir areas in Idaho that produce these bears serve to lessen the impact of high harvest rates on bears.

The harvest data we examined from the 1983–1990 mandatory check program showed that black bear populations in Idaho were experiencing moderate hunting pressure. In general, hunter numbers and harvest were increasing. We weren't sure, however, how these harvest rates were affecting populations because we didn't have precise data on harvest distribution or on population trends.

Our data showed that black bears in Idaho had low reproductive rates. We also know they are slow to respond to changes in harvest rates and that one- or two-week hunting season reductions don't affect harvests. As a result, we agree with Harris (2) that bears should be managed based on the risk of overharvest: If we wait until the scientific evidence of overharvest is undeniable, we will have to severely reduce season lengths and hunting opportunity. We also have an obligation to improve our ability to manage this species for future generations of Idahoans. Therefore, we recommend the following research and management actions:

1. Establish bait station surveys in key management areas (DAUs) to monitor black bear population needs.
2. Identify and protect important reservoir areas in each DAU.
3. Study subadult dispersal patterns, mortality rates for all age classes, and population responses to hunting regulation changes.
4. Improve data collection from harvested black bears to identify or validate criteria used for monitoring harvests.
5. Develop management objectives that will minimize both habitat loss and fragmentation (breakup), and ensure that protected movement corridors exist for dispersing subadults.
6. Prohibit the selling of black bear body parts. (There is a lucrative market in bear body parts. Allowing the sale of body parts contributes to increased illegal bear harvest.)
7. Adjust hunting season structures so more adult females will survive.
8. Work cooperatively with land management agencies to implement land use practices that will enhance the quality of black bear habitats in Idaho.

LITERATURE CITED

1. NELSON, L. J. 1991.
   1990 black bear opinion survey. Unpub. report, Idaho Dept. Fish and Game. 8pp.
2. HARRIS, R. B. 1984.
   Harvest age structure as an indicator of grizzly bear population status. M.S. thesis, University of Montana, Missoula. 204pp.
3. MILLER, S. D., AND S. M. MILLER. 1988.
   Interpretation of bear harvest data. Alaska Dept. Fish and Game, Fed. Aid in Wildl. Restor. Proj. W-22-6, Job 4. 18 R. 65pp.
4. BEECHAM, J. 1983.
   Population characteristics of black bears in west central Idaho. J. Wildl. Manage. 47(2):402–12.

5. MCILROY, C. W. 1972.

Effects of hunting on black bears in Prince William Sound. J. Wildl. Manage. 36:828–37.

6. RAYBOURNE, J. W. 1976.

A study of black bear populations in Virginia. Trans. Northeast. Sect., The Wildl. Soc., Fish and Wildl. Conf. 33:71–81.

7. LECOUNT, A. L. 1982.

Characteristics of a central Arizona black bear population. J. Wildl. Manage. 46:861–68.

8. WILLEY, C. H. 1974.

The Vermont black bear. Vermont Fish and Game Dept., Montpelier. 73pp.

9. POELKER, R. J., AND H. D. HARTWELL. 1973.

Black bear of Washington. State Game Dept. Biol. Bull. 14. 180pp.

10. CAUGHLEY, G. 1974.

Interpretation of age ratios. J. Wildl. Manage. 38:557–62.

11. MILLER, S. D. 1990.

Population management of bears in North America. Int. Conf. Bear Res. and Manage. 8:357–73.

ENDNOTES

[1] After a bear visits a bait set out by a houndsman, the houndman's dogs can pick up its scent.

ELEVEN | DENNING ACTIVITIES AND DEN CHARACTERISTICS

WHEN WE BEGAN OUR STUDIES OF BLACK BEARS IN Idaho during the summer of 1973, we didn't have much experience working with large, potentially dangerous animals. We often used trial and error to learn and refine many of our techniques for setting snares, using immobilizing drugs and radio-tracking bears. We also used trial and error when we began the daunting task of removing bears from their winter dens to replace radio collars and learn more about their denning habits. Prior to our studies, very few researchers had even attempted this task.

Although the work was logistically and physically demanding, it never lacked excitement. Each day we dealt with the danger of avalanches, the extremes of winter weather, and miles of backpacking our equipment on snowshoes. However, we also experienced the peaceful solitude of some of Idaho's most beautiful forests. We'll certainly miss that aspect of our denning work.

What we won't miss is the feeling we got when one of us crawled into a 15-by-20-inch hole to tranquilize an unrestrained bear. We couldn't see much detail or move around freely in the small, dark quarters, and we couldn't predict how the bear would respond to our intrusion. On one visit, we might encounter a bear who was so docile we could reach out and touch it before it was tranquilized. Another bear might charge and swat at us in the den entrance. We quickly learned to prepare for the worst behavior to avoid any serious problems for us or the bears.

During our denning work, bears bit crew members twice. Fortunately neither person was seriously hurt. The first instance occurred when a Norwegian colleague, Ivar Mysterud, was visiting. Ivar was studying European brown bears north of Oslo, but had never seen a wild bear in the forests of Norway. He was very excited about not only seeing a bear but also getting to handle it. On one occasion, Ivar had to help me remove a large female from her den. I had already tranquilized her, and Ivar was pulling on her front feet as I reached for her radio collar. Suddenly she whirled around, bit Ivar through the hand between his thumb and index finger, and then retreated back into the den. Needless to say, we gave her another dose of drugs before we made a second attempt to dislodge her from the den. Ivar took a full roll of photographs of his bloodied hand to show his family and friends back in Norway – his "war wound" was proof that he had actually handled bears in their dens while visiting the "wilds" of Idaho.

The second biting episode happened when my 12-year-old son Jay decided he was cold and lay down next to a tranquilized bear to nap while we finished taking den measurements. Before we were finished, the tranquilizer began to wear off. The bear awoke, moved her head to the side, and bit Jay on his lower leg. The bite barely broke the skin on Jay's calf, but it was enough to create another vivid memory.

JOHN BEECHAM

Idaho black bears spend half the year asleep in winter dens. Denning allows the bears to survive during a period when weather conditions are severe and food supplies are low. The denning habits of bears received little study prior to the widespread use of radiotelemetry equipment. Recently, many researchers have used radiotracking techniques to study the ecological and physiological properties of denned bears (1–11).

Our radio-telemetry studies of black bear population dynamics, activity patterns, movements, habitat use, and home range sizes (12–15) gave us the opportunity to locate and examine black bear dens. The objectives of our denning studies were to document the reproductive histories of radio-collared female black bears, to describe denning sites, and to replace radio collars. From the data we gathered, we were able to learn about denning chronology (when bears entered and emerged from their dens), and den types, locations, descriptions, selection, and use. We also looked at deaths that occurred in dens and the bears' responses to disturbance. Finally, we examined denning physiology (how the bears responded physically to denning), foot pad shedding during denning, and denning management considerations.

From 1973 through 1977, we used radio-telemetry techniques to locate and handle 49 black bears 83 times in 65 different dens at Council. We also found one den site during the 1975 summer field season. The sex, age, and reproductive status of denned bears by den type are presented in Table 11–1. Because of small sample sizes, we didn't include data from northern Idaho in this section.

Table 11–1. Den types used by black bears in west-central Idaho during 1973 through 1977 classified by bear sex, age, and reproductive class[a]

Den Type	Sex M	F	Age Subadult	Adult	Reproductive Class of Adult Females Single	Females w/Cubs	Females w/Yearlings	Other Single Bears
Ground	16	35	12	39	4	12	15	20
Tree[b]	7	6	4	9	2	2	2	7
Rock	–	2	–	2	1	1	–	–
Log	2	1	1	2	–	–	1	2
Total	25	44	17	52	7	15	18	29

[a] Dens that were reused are included.
[b] Den located in summer 1975 is not included.

DENNING CHRONOLOGY

The dates that bears entered their dens varied among bears and years (Table 11–2). Denning generally began in mid-October. Before entering their dens each fall, bears became less active. Females usually spent several days near their dens before actually entering them, and often acted lethargically. As an example, in 1975 bear No. U-100, an adult female, moved near her den on October 31 and stayed within a few hundred yards of her den site for several weeks. On November 12, we monitored her radio signal from about 200 yards away; for two hours she remained inactive. The following day we moved to within 30 yards of the den and saw her sitting lethargically outside the den opening. Seven days later she was inside her den and we were able to get within eight yards of the den without disturbing her.

Table 11–2. Range in dates that black bears entered their dens in west-central Idaho, 1973 through 1976. The number of bears studied is in parentheses

Year	Adult Males	Adult Females	Yearlings
1973	Oct 30–Nov 1 (2)	Oct 28–Nov 20 (5)	–
1974	Nov 11 (1)	Oct 9–Nov 8 (3)	Nov 27 (1)
1975	Oct 27–Nov 25 (8)	Nov 6–Nov 24 (7)	–
1976	Nov 2–Nov 7 (2)	Oct 15–Nov 8 (7)	Oct 10–Nov 16 (7)

During this waiting period, a few females evidently spent some time preparing for winter by gathering materials to line their dens. We observed only one female constructing a new den, but after a week she abandoned the effort for no apparent reason and denned in another location.

Males usually denned soon after arriving at their den site. In 1975, we located adult male No. U-18 2.4 miles away from his den three days before he denned. We also found that four days earlier, he had been 4.1 miles from his den, but only 1 mile from the den he selected the following year. The fact that males didn't linger near their dens before entering them was probably why we seldom observed the pronounced lethargic behavior in males that we saw in females. Rogers (16) reported similar behavior by males in Minnesota, and noted that males denned in the same small portion of their range each year. However, we found that Idaho males used dens that were scattered throughout their home ranges.

Our radio signals indicated that upon entering their dens the bears remained somewhat active, but this activity was probably limited to frequent adjustments to their body positions. If disturbed during this period, bears were likely to abandon their dens. During our study in west-central Idaho, we unintentionally disturbed 19 bears less than one week after they entered their dens. Eleven of the bears subsequently abandoned their dens. As the bears spent more time in their dens, their activity levels decreased, and the bears became less likely to leave their dens if disturbed.

Several factors affected the onset of denning: the availability of food, the females' reproductive condition, and the weather. The relative importance of these factors in different geographic locations varied from year to year throughout the bears' range. It appeared as well that the sex, age, or reproductive condition of bears may have influenced how they responded to these factors. These differences probably accounted for much of the variation we observed in the denning studies.

Between 1975 and 1976 (the study years with the greatest contrast in weather at Council), the average dates bears entered their dens differed by 15 days (November 14, 1975, and October 30, 1976), a fact we attribute to the availability of food. The bears' preferred food plants matured about two weeks later in 1975 than in 1976. As a result, in 1975 some bears kept feeding even though several inches of snow had accumulated. In 1976, when foods were scarce, all radio-tagged bears denned earlier, even though not much snow had fallen and daily temperatures were unusually mild (the mean daily maximum temperature during the period bears entered their dens was 48°F in 1975 and 61°F in 1976).

Pregnancy also influenced the onset of denning. In 1975, when most of the Council females we studied weren't pregnant, males entered their dens an average of eight days before females.

The following year, when all the females were pregnant, the females denned ten days earlier than the males. Although small
sample sizes and high variability in the onset of denning made interpretation of the data difficult, we believe that pregnancy was the
reason females denned earlier than males in 1976. Lindzey and
Meslow (5) reported that pregnant females denned before males,
but didn't mention nonpregnant females. Erickson (17) and Johnson and Pelton (18) reported that female and juvenile black bears
denned before adult males in Michigan and Tennessee. Pearson
(19) stated that adult male grizzly bears remained active longer
than females and younger bears.

Weather didn't seem to greatly affect the onset of denning.
From October 26 to November 26, 1975, most bears entered their
dens on days when the daily maximum temperature was $45°F$
rather than the average $48°F$. However, in 1976 we saw no apparent
relationship between temperature and denning over the 34-day period bears entered dens. In 1975 and 1976, we noted no significant
correlation between precipitation and the date bears entered dens.
In 1976, it rained for only two days (for a total rainfall of 0.21 inches)
during the period bears were entering their dens.

The onset of denning by bears in this study was comparable to
that reported by Erickson (17, 20) in Michigan and Alaska, Jonkel
and Cowan (21) in Montana, and Poelker and Hartwell (22) and
Lindzey and Meslow (5) in Washington. Rogers (16) reported that
some bears in Minnesota denned as early as September 2. Johnson
and Pelton (18) reported that *circa-annual* rhythms synchronized the
denning behavior of black bears with the environment. Mysterud
(23) described winter denning in bears as an elaborate bedding process that evolved as a result of adverse environmental conditions in
the Palearctic region. He suggested that four factors were important in the evolution of bedding systems. They included
(1) concealment, (2) exposure, (3) construction constraints, and
(4) defensive/psychological factors. The relative importance of
these factors probably varied from area to area, depending on the
bear species, climate, geology, and the bear's sex and age.

On the other end of the denning period, most bears began

leaving their dens in mid-April. We obtained data on when bears emerged from their dens for only two years: 1974 and 1976. In 1974 bears left their dens from April 11 to April 30; two females accompanied by yearlings were the last to emerge (13). In 1976, one adult male left his den on March 21 because we disturbed him. Two adult males and one female with yearlings emerged between March 23 and April 10. Eight adult bears (three adult males and five females with yearlings) left their dens between April 10 and May 8.

Although our data weren't sufficient to detect any differences in the times bears emerged from their dens, our observations suggested that females with newborn cubs were the last to leave their dens. In addition, bears who denned at lower elevations generally emerged before those who denned at higher elevations. The emergence dates in this study were comparable to dates reported for Maine (24), Alaska (20), North Carolina (10), and Minnesota (16). Lindzey and Meslow (5) and LeCount (11) reported earlier emergence dates for western Washington and Arizona. Jonkel and Cowan (21) reported slightly later emergence dates for northwestern Montana.

Even though bears left the den site soon after emerging, they didn't immediately become fully active. In most cases, bears moved to areas below the snow line, where they constructed *day beds* by scraping vegetation into a pile at the base of a tree. The bears moved varying distances from their dens after emerging, depending partly on the distance to the nearest available food and (for females) whether they had cubs. They usually remained close to their day beds for several days and abandoned them only when their activity levels increased. Rogers (16) reported similar activities for black bears in Minnesota.

DEN TYPES AND LOCATIONS

The sex, age, and reproductive class of Idaho black bears didn't influence the type of den they selected (Table 11–1). Forty-seven (71 percent) of the 66 dens we located during the Council study were ground dens dug into a hillside or under the base of a tree, stump, shrub, or fallen tree (Plate 9). We found 14 (21 percent) of the dens in the base of hollow trees and five (8 percent) of the dens in hollow logs or rock cavities. We saw several bears use both excavated ground dens and tree dens. During our study, female bear No. U-41 used six ground dens, one tree den, and one log den.

Jonkel and Cowan (21) and Lindzey and Meslow (6) reported that black bears denned primarily in hollow trees in Montana and Washington. Erickson (17), however, found that Michigan bears used excavated dens more frequently than other den types. Black bears occasionally used man-made structures for denning, such as drainage culverts (25) and deserted buildings (17, 21). In contrast, grizzly bears usually dug ground dens (1, 2, 26, 19, 27, 7, 28, 29, 30).

Grand fir was the most common tree species used by black bears for denning. Eight bears used grand fir, three used subalpine fir, and one used Douglas fir. Eighty-six percent of the tree dens were located in the base of live trees. We found tree dens at significantly higher elevations than ground dens, probably because of the way trees were distributed on the study area. Ponderosa pine, the dominant species at lower elevations, wasn't used for denning, while grand fir and subalpine fir, found at higher elevations, were both suitable species for denning.

Black bears denned at all elevations, on all slopes, and under a variety of canopy coverages (Table 11–3). However, we noted that bears preferred elevations and aspects that provided cover. For example, adult male No. U-18 used three dens: one located at 7,135 feet in a dense subalpine fir stand and two located at 4,100 feet. We found one of his lower-elevation dens in a brush pocket over 100 yards from the nearest stand of timber. Dens at lower elevations (less than 4,100 feet) were often located in dense pockets of brush, primarily ninebark, while dens above 5,900 feet were often found in relatively open areas. Heavy snow accumulation probably provided adequate hiding cover for bears denning at high elevations, but bears denning at lower elevations needed the additional protection of thick brush.

Aspect also influenced hiding cover. We located 37 (59 percent) of 62 dens on west, northwest, and north aspects (Fig. 11–1). In the Council area, low and middle-elevation timber stands were

Table 11–3. Mean elevation, slope, and canopy coverage at black bear dens in west-central Idaho, 1973 through 1977

	Elevation (feet)		Slope (°)			Canopy Coverage (%)		
No. Dens[a]	Mean Elev.	Range in Elev.	No. Dens	Mean Slope	Range in Slope	No. Dens	Mean Can.	Range in Can.
Sex								
Adult male	10	5,118	3,904–6,808	10	20	6–31	10	43	0–95
Adult female	37	5,004	3,609–6,644	34	22	7–47	36	37	0–95
Den Type								
Ground	46	4,757	3,511–6,808	42	23	6–47	44	44	0–97
Tree	14	5,955	5,102–6,431	14	16	6–25	14	41	0–95

[a] Number of dens studied.

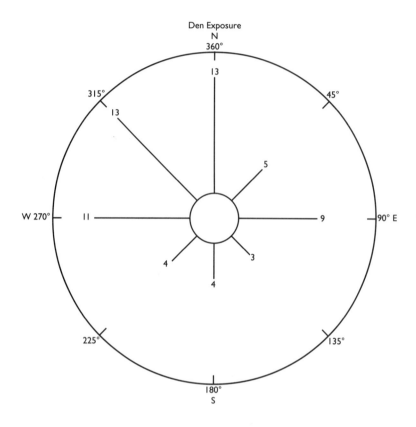

Figure 11–1. Aspects of 62 black bear dens at Council, 1973–77.

found primarily on northwest and north aspects. These shady exposures usually allowed snow to accumulate and persist for long periods, resulting in favorable soil moisture conditions for growing the thick brush stands that provided bears with hiding cover. In alpine areas, bear dens were often found on the leeward side of windblown ridges where deep snowdrifts occurred. We found no significant differences in the aspects used by adult males and females, or those used by single bears and females accompanied by young.

Lindzey and Meslow (6) reported that aspect didn't influence den site selection for black bears in Washington. Craighead and Craighead (1, 2) and Lentfer et al. (26) indicated that grizzly bears in

Yellowstone National Park and Kodiak Island, Alaska, selected north aspects, while Troyer and Faro (27) and Reynolds et al. (7) found that grizzlies denned most often on south aspects in other areas of Alaska. Pearson (19) found no bears denned on north aspects in the Yukon Territory, but bears used all other aspects equally. In northern latitudes, permafrost may have prevented bears from excavating dens on north slopes.

Thirty-one (52 percent) of the 60 black bears we studied denned on slopes of 20 to 40 degrees. Although similar data aren't available for black bears in other areas, most grizzlies studied by researchers denned on slopes of 20 to 40 degrees (26, 19, 27, 7, 28, 29, 30).

Most bears chose denning sites away from roads or human activities, but there were exceptions. Several adult bears at Council denned less than 50 yards from well-used roads. Two of bear No. U-41's eight dens were located less than 10 yards from open roads, and in 1975 she produced two cubs in one of these dens. Adult females often denned in relatively open clear-cuts. Lindzey and Meslow (6) found that in western Washington, yearling black bears tended to den in recently cut areas, while adults denned primarily in timbered areas. They suggested that adult bears selected the most secure areas for denning. Craighead and Craighead (1, 2) reported that grizzly dens in Yellowstone National Park were located considerable distances from developed areas or human activity.

DEN DESCRIPTIONS

During this study, we measured the entrance height and width; tunnel height, width, and length; and chamber height, width, and depth of 47 ground and 13 tree dens (Table 11–4, Fig. 11–2). We saw no significant differences in den dimensions for specific sex or age classes except that adult males dug larger den entrances than did other bears. Although adult male black bears were larger than females, females with young needed den chambers as big as those required by adult males. However, females with young didn't need large entrances. These data suggested that bears dug entrances just large enough to accommodate their body size.

Data from other researchers comparing the physical dimen-

Table 11–4. Dimensions of excavated (E) and tree dens used by black bears in west-central Idaho, 1973 through 1977

	Mean Entrance (in)		Mean Tunnel (in)			Mean Cavity (in)		
	Height	Width	Height	Width	Length	Height	Width	Depth
Den Type								
Excavated	17.5	22.4	16.5	24.9	38.0	28.2	40.9	43.8
Tree	15.5	14.3	–	–	88.9	37.9	35.2	
Reproductive Class								
Female w/Cubs (E)	19.4	20.2	17.0	25.0	17.2	33.0	41.3	43.0
Female w/Yearlings (E)	15.8	20.2	14.9	23.7	55.1	28.9	44.2	44.0
Single Bears (E)	17.8	25.1	16.9	25.4	39.7	25.6	38.5	43.9
Sex								
Adult Male (E)	21.4	28.2	18.0	27.0	19.3	27.2	40.8	46.7
Adult Female (E)	17.2	16.2	15.9	24.8	26.3	30.6	42.1	42.5

Figure 11–2. The mean diameter of ten tree dens we examined was 44 inches. Photo by John J. Beecham

sions of black bear dens by sex, age, or reproductive class of black bears are lacking. Lindzey and Meslow (6) examined 11 black bear dens in Washington and observed no difference in size between those of adult females and yearlings. Craighead and Craighead (1, 2), however, reported that female grizzlies with young dug wider dens than bears who denned alone.

Twenty (45 percent) of 44 ground dens we examined had tunnels leading from the entrance to the den chamber. Two dens contained two den chambers, and three dens had two entrances. Eight (17 percent) of 47 ground dens we looked at had small tunnels averaging 8.7 inches in height and 12.4 inches in width leading away from the main chamber. We believe that bears enlarged portions of old coyote or badger dens for their use and these small tunnels were remnants left by the previous occupants. Matson (31) reported that a female black bear in Pennsylvania apparently renovated an old woodchuck or fox den for herself.

Other research showed that the physical size of bear dens varied, depending on the type of den used, the den material selected, and the species of bear involved. Pearson (19) suggested that the most important factors regulating the size of excavated grizzly bear dens in the Yukon Territory were the size of the bear and the amount of air space within the den that must be warmed by the bear. However, Craighead et al. (32) found that denned bears lived in a relatively confined environment, and reported that den size wasn't important in regulating den temperature.

Thirty-two (76 percent) of 42 excavated dens we examined contained nest materials. We found no significant differences between the proportion of males and females who lined their dens with nest materials. Erickson (17), Jonkel and Cowan (21), and Lindzey and Meslow (6) reported that 30 percent, 30 percent, and 80 percent, respectively, of the dens they examined contained nest materials. Erickson (17) suggested that fewer adult males lined their dens than did females or juvenile bears.

The bears constructed most of their nests with brush or conifer boughs dragged into the den from the area immediately surrounding the den. Pregnant females tended to use grass as a nest material more often than single bears or females accompanied by yearlings. In log and tree dens, bears used woodchips most often as nest materials; only one tree den contained materials brought in from the outside. Jonkel and Cowan (21) observed that some bears blocked the entrance to their den with nest materials. We saw this behavior most commonly at tree dens.

For some bears, nest construction behavior varied from year to year. Bear No. U-41 lined five of the eight dens she used during the study. However, two of her unlined dens were located in a tree or log – den types that usually contained no nest materials except woodchips removed from the inside walls of the den.

DEN SELECTION AND USE

Black bears in west-central Idaho apparently maintained several dens within their home ranges for hibernation. Adult female No. U-41 occupied at least eight different dens during five winters; several other bears occupied three and four different dens during the years they were radio-collared.

We found that bears will reuse their dens each year if they're not disturbed. Two bears that we disturbed did reuse their dens, but not in consecutive years.

The tendency of bears to reuse traditional sites varied considerably from area to area, depending on the durability of the dens. Craighead and Craighead (1, 2) and Reynolds et al. (7) observed very few bears reusing their dens in Yellowstone National Park and northeastern Alaska. They attributed this to the tendency of these dens to collapse during the spring thaw. Other investigators saw black and grizzly bears reuse their dens (21, 26, 19).

Lindzey and Meslow (6) suggested that the tendency of bears to use previously constructed dens may be reflected by the similarity in chamber size and entrance height of dens occupied by adult females and their yearlings. We believe that yearlings, especially females, often used dens constructed and maintained by their mothers. Although our radio-tracking data indicated that the bears didn't use their dens extensively during the summer, we thought they visited them periodically, allowing their offspring to learn the den locations during these visits. Servheen and Klaver (30) found that grizzly bears in northwestern Montana visited their dens occasionally during the summer months.

Two of 39 dens we examined in late November 1976 and early December 1977 were occupied by the offspring of the females who originally used the dens. In one case, a 22-month-old female, No. U-130, used a den that her mother had occupied in 1974, a year before her birth. The den hadn't been used since 1974, suggesting that the adult female had visited the den site sometime during the previous summer to show No. U-130 its location.

However, not all yearlings used dens built by their mothers. Of the six yearlings studied, two reused their mothers' dens, while two constructed new dens for their first denning efforts. One yearling attempted to build his own den both under and within the same log, but didn't finish it. Another yearling male left the study area and denned in a grand fir 13.7 miles south of his mother's home range.

These data and other evidence suggested that bears didn't need their mothers to learn how to den. During this study, we released several cubs and yearlings who had been held in captivity for 3 to 18 months (33, 34). Later, we recaptured several of these bears who had successfully denned without learning denning behavior from their mothers when they were yearlings. One male, No. U-71, was captured as a cub, held in captivity for 18 months, and then released on November 10, 1974, when there were about 10 inches of snow on the ground. We recaptured him on our study area three times the following summer. Our data indicated that denning behavior was instinctive rather than learned, so juvenile bears didn't need extensive parental care after weaning to survive the winter period.

We saw only family groups sharing dens. Lentfer et al. (26) observed one instance of unrelated bears sharing a den site.

DEN-ASSOCIATED MORTALITY

We found three black bears dead in their dens during this study, but we believe we accidentally caused those deaths. During the early years of this study, we returned all bears handled at the den sites to their dens before they recovered from the immobilizing drug. The bears probably suffocated when their tongues slid into the back of their mouths, blocking air passage. When we delayed returning the bears to their dens until they had regained reasonable control over their head movements, no additional deaths occurred. We saw no natural deaths during the denning period.

Observations of den-associated deaths are extremely rare. Reynolds et al. (7) saw two examples of den-related mortalities in Alaskan grizzlies, and Pearson (19) reported one instance in his Yukon Territory study. Jonkel and Cowan (21) observed no deaths of denned black bears in Montana, nor did Lentfer et al. (26) for grizzlies on Kodiak Island and the Alaskan peninsula.

Craighead and Craighead (1, 2) suggested that old bears may die in or near their dens. Our observations on the physical condition of denned black bears showed that they remained in good condition throughout the denning period. If any physical stress occurred, it was during early July, when their fat reserves were poorest. We saw one example of natural mortality: An 18-year-old female died in mid-July.

RESPONSE TO DISTURBANCE

During our studies, we noted how bears reacted when we disturbed them in their dens. As we approached, the bears responded in a variety of ways. All bears acknowledged our presence at the den by raising their heads and looking at us, but only rarely did they become aggressive. Generally, the bears reacted lethargically to our activities around the den. The most common response was that the bear faced the den entrance and smelled the jab stick as we poked it into the den. Occasionally a bear attempted to bite the jab stick – moving almost in slow motion – but most often they just smelled it.

We saw two forms of aggression. In the first form, the bear would vocalize by "chomping" its jaws, "blowing," or growling. In the second form, the bear rushed toward the den entrance and swatted its forepaws at us or the jab stick while we tried to immobilize it. Normally, we didn't see this behavior unless we were actively attempting to inject the immobilizing drug into the bear. Those rare instances of aggressiveness weren't characteristic of any particular age, sex, or reproductive class. They were only predictable in the sense that they tended to occur more often in the early and very late stages of the denning period, when the bears were less sleepy.

Several bears abandoned their dens after we disturbed them during the early stages of the denning period, but only two bears left their dens when disturbed in late winter (Fig. 11–3). Bear No. U-49, an adult male, was particularly sensitive to disturbances near his den. We were never able to successfully approach one of his dens without being detected, causing him to abandon the den.

Lindzey and Meslow (5) reported that black bears tended to stay in their dens after entering them, but didn't comment on how bears responded to a researcher's approach. Jonkel and Cowan (21) reported that bears were less sensitive to their presence; only 2 of 40

Figure 11–3. When one bear abandoned its den during our denning studies, we captured it using trailing hounds and backpacked the bear to its den site. Photo by Ivar Mysterud

bears they observed during the denning period showed considerable antagonistic behavior. Craighead and Craighead (1, 2) found that a female grizzly accompanied by two grown cubs was aroused by their presence, but was so sleepy that she didn't attempt to leave the den. Other researchers reported similar responses for black bear females accompanied by young (35, 31).

PHYSIOLOGY OF DENNED BEARS

Physiological data collected from denned bears and observations of their behavior immediately before and after the denning period suggested that they experienced a metabolic shift prior to denning that made them become more lethargic as the denning period progressed. The metabolic shift was apparently reversed sometime before the bears left their dens, but the bears didn't become fully active until several weeks after emergence.

There is some uncertainty about the true physiological status of denned bears. Although the bears' heart rates and respiration decline significantly, as they do in other hibernators, their body temperatures drop only slightly (36, 37, 38, 2, 39, 4). Hock (40) proposed that the term "carnivorean lethargy" be used to describe the physiological condition of black bears during the denning period, while Morrison (41) suggested they be called "ecological" rather than physiological hibernators. Folk et al. (4) suggested that bears were more efficient hibernators than most animals exhibiting the hibernation phenomenon. We believe that black bears are more efficient hibernators than smaller mammals because they don't awake during the winter to urinate, defecate, or feed, as many "true" hibernators do every three to ten days.

The black bears we studied in west-central Idaho had mean rectal temperatures of 96.8°F after we immobilized and removed them from their dens. Our data (Table 11–5) suggested that their body temperatures dropped as the period of dormancy progressed and then rose again before they left the den. Craighead et al. (32) reported that the *telemetered rectal temperature* of a black bear in February was about 95.4°F. They also stated that the mean temperature of this bear was slightly higher in March, but suggested that disturb-

Table 11–5. Mean respiration, temperature, and heart rates of active and dormant black bears captured in west-central Idaho, 1973 through 1977

Month	Respiration (bpm)	Temperature (°F)	Heart Rate (bpm)
Active			
May	12.6	100.4	139.9
June	14.0	100.8	127.3
July	16.8	101.3	131.9
August	19.0	101.5	142.8
September	12.0	101.7	131.0
October	13.5	100.8	137.9
Dormant			
November	5.9	96.6	87.9
December	7.5	94.3	93.7
March	7.2	97.9	89.1

ing the bear a second time caused the higher recording. Nelson et al. (39) reported that the winter rectal temperatures of three captive bears were between 93° and 95°F and their *respiratory quotients* decreased gradually during winter and rose more rapidly in spring after the bears were active and awake.

FOOT PAD SHEDDING

Some researchers (42) have observed that bears shed their foot pads while denning. We examined the foot pads of 49 different bears on 83 occasions during winter dormancy (13 in November, 13 in December, and 56 in March). We saw no instances of foot pad shedding except during March. Of the 56 bears we handled in March, three had completely shed their foot pads, three showed no apparent shedding, and the remainder were in various shedding stages. The only apparent pattern in the shedding sequence was that bears often shed their digital pads before their plantar pads. In addition, the front pads were usually more completely shed at the time of examination than the rear plantar pads. Newborn cubs didn't shed their foot pads, but bears of all other ages did.

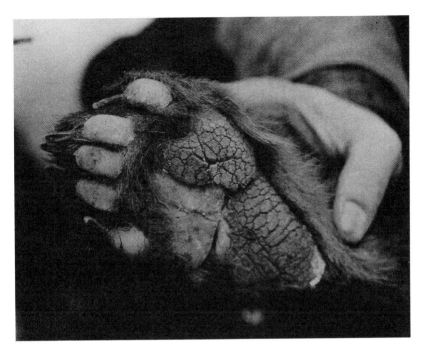

During denning, some of the bears we studied shed their foot pads. Most of the foot pad shedding occurred in March. Photo by John J. Beecham

Rogers (42) described the shedding of foot pads by black bears during their denning period in Minnesota. He suggested that tender paws may be a factor affecting the early spring movements of black bears. We agree that foot pad shedding may have hampered bear movements immediately after they emerged from their dens, but we believe the bears' physiological condition (being lethargic) restricted their movements more than any other factor.

MANAGEMENT CONSIDERATIONS

Although extensive logging operations on the Council study area certainly reduced the number of suitable trees for denning, bears could easily dig ground dens in the area's soils. Most radio-collared bears maintained and occupied several dens within their respective home ranges. We concluded that the availability of dens wasn't limiting bear population size in west-central Idaho and that most black bears at Council didn't require large areas of undisturbed habitat for denning. However, in areas where soil conditions aren't suitable for excavating dens, black bears may be dependent on old-growth timber stands for denning sites.

Bears did prefer concealed den sites, as reflected in the tendency of bears denning at lower elevations to select thick brush pockets for den sites, while those denning at higher elevations often chose relatively open areas where snowpacks provided cover. Avoiding exposure to inclement weather was also an important consideration at Council; bears most often denned on aspects where snow persisted for long periods. Our lack of data on the availability of different aspects, however, kept us from drawing any definitive conclusions from these observations.

The dens we examined weren't concentrated in any core areas either within the home range of an individual bear or in the study area. As a result, it wasn't feasible to manage specific locations in the Council study area as core denning areas, although this could be done in areas where bears extensively use trees for denning.

LITERATURE CITED

1. CRAIGHEAD, J. J., AND F. C. CRAIGHEAD, JR. 1972A.
Grizzly bear prehibernation and denning activities as determined by radiotracking. Wildl. Monogr. No. 32. 35pp.

2. CRAIGHEAD, J. J., AND F. C. CRAIGHEAD, JR. 1972B.
Data on grizzly bear denning activities and behavior obtained by using wildlife telemetry. Int. Conf. Bear Res. and Manage. 2:84–106.

3. CRAIGHEAD, J. J., J. R. VARNEY, F. C. CRAIGHEAD, JR., AND J. S. SUMNER. 1976.
Telemetry experiments with a hibernating black bear. Int. Conf. Bear Res. and Manage. 3:357–71.

4. FOLK, G. E., JR., A. LARSON, AND M. A. FOLK. 1976.
Physiology of hibernating bears. Int. Conf. Bear Res. and Manage. 3:373–80.

5. LINDZEY, F. G., AND E. C. MESLOW. 1976A.
Winter dormancy in black bears in southwestern Washington. J. Wildl. Manage. 40:408–15.

6. LINDZEY, F. G., AND E. C. MESLOW. 1976B.
Characteristics of black bear dens on Long Island, Washington. Northwest Sci. 50:236–42.

7. REYNOLDS, H. V., J. A. CURATOLO, AND R. QUIMBY. 1976
Denning ecology of grizzly bears in northeastern Alaska. Int. Conf. Bear Res. and Manage. 3:403–409.

8. JOHNSON, K. G. 1978.
Den ecology of black bears in the Great Smoky Mountains National Park. M.S. thesis, University of Tennessee, Knoxville. 107pp.

9. PELTON, M. R., L. E. BEEMAN, AND D. C. EAGAR. 1980.
Den selection by black bears in the Great Smoky Mountains National Park. Int. Conf. Bear Res. and Manage. 4:149–51.

10. HAMILTON, R. J., AND R. L. MARCHINTON. 1980.
Denning activity of black bears in the coastal plain of North Carolina. Pages 121–126. In C. J. Martinka and K. L. McArthur, eds. Bears – Their biology and management. 4:121–26.

11. LECOUNT, A. L. 1980.
Some aspects of black bear ecology in the Arizona chaparral. Int. Conf. Bear Manage. and Res. 4:175–79.

12. BEECHAM, J. 1983.
Population characteristics of black bears in west central Idaho. J. Wildl. Manage. 47(2):402–412.

13. AMSTRUP, S. C., AND J. J. BEECHAM. 1976.
Activity patterns of radio-collared black bears in Idaho. J. Wildl. Manage. 40: 340–48.

14. REYNOLDS, D. G., AND J. J. BEECHAM. 1980.
Home range activities and reproduction of black bears in west central Idaho. Int. Conf. Bear Res. and Manage. 4.

15. YOUNG, D. D., AND J. J. BEECHAM. 1986.
Black bear habitat use at Priest Lake, Idaho. Int. Conf. Bear Res. and Manage. 6:73–80.

16. ROGERS, L. L. 1977.
Social relationships, movements, and population dynamics of black bears in northeastern Minnesota. Ph.D. thesis, University of Minnesota, Minneapolis. 194pp.

17. ERICKSON, A. W. 1964.
A mixed-age litter of brown bear cubs. J. Mammal. 45(2):312–13.

18. JOHNSON, K. G., AND M. R. PELTON. 1980.
Environmental relationships and the denning period of black bears in Tennessee. J. Mammal. 61:653–60.

19. PEARSON, A. M. 1975.
The northern interior grizzly bear *Ursus arctos*. Can. Wildl. Serv. Rep. Ser. No. 34, Ottawa. 86pp.

20. ERICKSON, A. W. 1965.
The brown-grizzly bear in Alaska. Its ecology and management. Fed. Aid Wildl. Restor. Proj. W-6-R-5, Work Plan F. Vol 5. Alaska Dept. Fish and Game, Juneau. 42pp.

21. JONKEL, C. J., AND I. M. COWAN. 1971.
The black bear in the spruce-fir forest. Wildl. Monogr. No. 27. 55pp.

22. POELKER, R. J., AND H. D. HARTWELL. 1973.
Black bear of Washington. State Game Dept. Biol. Bull. 14. 180pp.

23. MYSTERUD, I. 1987.
Bedding behavior in the European brown bear. *In* E. C. Meslow, ed. Bears – Their biology and management.

24. SPENCER, H. E., JR. 1955.
The black bear and its status in Maine. Maine Dept. Inland Fish and Game Bull. 4. 55pp.

25. BARNES, V. G., JR., AND O. E. BRAY 1966.
Black bears use drainage culverts for winter dens. J. Mammal. 47:712–13.

26. LENTFER, J. W., R. J. HENSEL, L. H. MILLER, L. P. GLENN, AND V. D. BERNS. 1972.
Remarks on denning habits of Alaska brown bears. Int. Conf. Bear Res. and Manage. 2:125–32.

27. TROYER, W., AND J. B. FARO. 1975.
Aerial survey of brown bear denning in the Katmai area of Alaska.

Pres. at Northwest Sect. Wildl. Soc. Meet. 2–4 Apr. 1975, Anchorage, Alas. 10pp.

28. RUSSELL, R. H., J. W. NOLAN, N. G. WOODY, G. H. ANDERSON, AND A. M. PEARSON. 1978.
A study of the grizzly bear in Jasper National Park: A progress report 1976 and 1977. Prep. for Parks Canada. Prep. by Can. Wildl. Serv., Edmonton, Alberta. 95pp.

29. VROOM, G. W., S. HERRERO, AND R. T. OGILVIE. 1980.
The ecology of winter den sites of grizzly bears in Banff National Park, Alberta. Int. Conf. Bear Res. and Manage. 4:321–30.

30. SERVHEEN, C., AND R. KLAVER. 1983.
Grizzly bear dens and denning activity in the Mission and Rattlesnake mountains, Montana. Int. Conf. Bear Res. and Manage. 5:203–209.

31. MATSON, J. R. 1954.
Observations on the dormant phase of a female black bear. J. Mammal. 35(1):28–35.

32. CRAIGHEAD, J. J., F. C. CRAIGHEAD, JR., J. R. VARNEY, AND C. E. COTE. 1971.
Satellite monitoring of black bear. BioScience. 21(24):1206–11.

33. JONKEL, C. J., C. LEE, T. THIER, N. MCMURRAY, AND R. MACE. 1979.
Sullivan Creek grizzlies. Border Grizzly Proj., Univ. Mont. Spec. Rep. No. 23. 10pp.

34. ALT, G. L., AND J. J. BEECHAM. 1984.
Reintroduction of orphaned black bear cubs into the wild. Wildl. Soc. Bull. 12:169–74.

35. ALDOUS, S. E. 1937.
A hibernating black bear with cubs. J. Mammal. 18(4):466–68.

36. MATSON, J. R. 1946.
Notes on dormancy in black bear. J. Mammal. 27(3):203–12.

37. HOCK, R. J. 1951.
Rectal temperatures of the black bear during its hibernation. Proc. Second Alas. Sci. Conf. 2:310–12.

38. SVIHLA, A. H., AND H. S. BOWMAN. 1954.
Hibernation in the American black bear. Am. Mid. Nat. 52(1):248–52.

39. NELSON, R. A., N. W. WAHNER, J. D. JONES, R. D. ELLEFSON, AND P. E. ZOLLMAN. 1973.
Metabolism of bears before, during, and after winter sleep. Am. J. Physiol. 224(2):491–96.

40. HOCK, R. J. 1960.
Seasonal variation in physiological functions of Arctic ground squirrels and black bears. Pages 155–171. In C. P. Lyman and A. R. Dawe, eds. Mammalian hibernation I. Harvard University, Cambridge, Mass.

41. MORRISON, P. 1960.
Some interrelations between weight and hibernation function. Bull. Mus. Comp. Zool. Harv. 124:75–91.

42. ROGERS, L. L. 1974.
Shedding of foot pads by black bears during denning. J. Mammal. 55(3):672–74.

THE BLACK BEAR IS A SHY ANIMAL WIDELY DISTRIBUTED in Idaho. North of the Snake River plain, bears inhabit the state's forested mountains and foothills; few bears live south of the Snake River, except in southeastern Idaho. Even though Idaho black bears don't occur in high densities, the bear is popular with hunters, outfitters, and wildlife watchers.

The bear family originated in the Palearctic region more than 20 million years ago. During the Pleistocene era, three descendants of the ursid family – the polar, black, and brown bears – moved to North America. Although its range has shrunk in recent years, the black bear is still found in all or parts of 38 states, 11 Canadian provinces, and 7 Mexican states. Most of Idaho's bears are found in the state's coniferous forests. We calculated that Idaho's 30,000 square miles of bear habitat could support a maximum of 20,000 to 25,000 bears. Our studies indicate that the actual number of Idaho bears is probably less than this.

Since the early 1970s, we have been collecting the biological data needed to manage bears. Between 1972 and 1990, we studied bear populations in six areas: near Council in west-central Idaho; near Lowell in north-central Idaho; the Coeur d'Alene River drainage north of Wallace; the east side of Priest Lake; the Elk River drainage; and the St. Joe River drainage. All of these areas have continental climates that are influenced by air currents from the Pacific Ocean, producing long, cold, wet winters and dry, hot summers.

The Council study area is drier than the northern Idaho study areas, but contains productive vegetation: sagebrush/grasslands at lower elevations and on drier sites, and timber stands at higher elevations and on wetter sites. The Council study area contains many naturally occurring open areas interspersed throughout the timbered areas. On the northern Idaho study areas, trees dominate the vegetative communities. The Lowell, Coeur d'Alene and St. Joe study areas also have large seral brushfields created by wildfires.

Cattle grazing and logging are the major land uses on the Council and Elk River study areas. Timber production is important on the Coeur d'Alene, Priest Lake, and St. Joe areas. Moderately to heavily logged study areas contain many roads and are accessible to the public. The Lowell study area, in contrast, isn't logged and is accessible only by one road.

During these studies, we looked at the status of each bear population and collected data on the bears' food habits, physical condition, denning requirements, behavior, and habitat use patterns. We also developed a system for monitoring bear population trends.

To get an accurate picture of black bear life, we trapped bears, marked and/or radio-collared them, released them, and later recaptured them. Bears were lured to cubby sets with baits, captured in foot snares, and injected with an immobilizing drug. While the bears were unconscious, we inserted ear tags and tattooed an identifying number on their ear and upper lip. We also took physical measurements, blood samples, and other physiological data from the bears. On the Council and Priest Lake study areas, we placed radio collars around some of the captured bears' necks. We later monitored those bears from the ground and from an airplane.

To learn what types of habitats bears used, we periodically located the bears using radio transmitters and described the areas where we found them. Bear habitats were classified by the type of plants found, the amount of horizontal and vertical cover those plants provided, and the topography of the area.

We located bear dens by monitoring radio-collared bears during October and November. In December or March, we shoveled snow from the den entrances, immobilized the bears, and removed them from the dens. Most of the bears were weighed and measured, and the physical characteristics of the dens and den sites were recorded. Before we left, we removed or replaced old transmitter collars, returned the bears to their dens, and covered the den entrances with snow.

We used two methods to obtain data on bears killed by hunters. The mandatory check and report program required a hunter to bring the skull of his/her harvested bear to an official checkpoint within ten days of killing the bear. We extracted a tooth from the skull to age the bear and recorded information about the kill. Each year, we also contacted about 3 percent of bear tag holders with a telephone survey to estimate the number of bears harvested and hunter success.

MORPHOLOGY AND PHYSIOLOGY

Our studies showed that black bears grew rapidly until puberty, and then grew more slowly as each individual approached his or her maximum size. Genetic differences affected bear size: A small mother produced small cubs. During the summer, Idaho male bears generally weighed 250 to 280 pounds, while females weighed 120 to 140 pounds. On average, adult males were 77 pounds heavier and 6 inches longer than adult females. Bear weights varied seasonally; they gained weight in late summer/fall, when they ate berries and other nutritious foods, and lost weight during the hibernation period and during late spring and early summer. Nursing females, who lost large amounts of body fat while lactating, showed especially large weight fluctuations.

Although growth rates were essentially the same for all the populations we studied, Council bears grew larger than did those from other study areas. This suggested that nutritional rather than genetic differences caused the heavier weights. The quantity and quality of available foods also affected the bears' age of first reproduction, litter size, and litter frequency. Our research showed that Idaho females needed to reach a body weight of about 100 pounds before they produced young. This often occurred when the bears were four to seven years old. None of the bear populations we studied achieved their maximum growth or reproductive potential. Because nutrition was the primary factor affecting those rates, we believe that enhancing bear habitats could make bear populations more productive.

North American black bears have several color phases. Most of the Council bears we captured were brown, whereas the majority of the northern Idaho bears we trapped were black. We believe that brown-phase bears were better able to tolerate hot, dry, sunny habitats (such as Council) than were black-phase bears. In more humid, less sunny, northern Idaho habitats, brown-phase bears had no advantage.

Our tests of blood samples from Council and Lowell bears showed that the most prevalent diseases that bears were exposed to included tularemia, brucellosis, and toxoplasmosis. Some bears tested positive for more than one disease. We observed no clinical evidence of disease in black bears and didn't believe diseases were a major concern from a population perspective during our studies.

The external parasites we found most often on black bears were ticks, lice, and fleas. Although ticks were most commonly found, most bears had fewer than 25. We saw lice and fleas infrequently. Our blood tests also showed that bears were exposed to trichinosis, an internal parasite.

The pesticide and mercury levels we measured in bears were well below the maximum tolerances allowed in domestic livestock by the U.S. Food and Drug Administration. We don't believe these toxins had a significant effect on black bear survival.

BEHAVIOR

Our study of bear activity patterns showed that Council and Priest Lake bears were most active between 5:00 a.m. and 10:00 a.m. and between 6:00 p.m. and 9:00 p.m. They were least active between 1:00 a.m. and 4:00 a.m. Bears were less active in the spring, when food supplies were of marginal quality, and more active in the summer, when energy-rich, nutritious foods were available. Bear activity levels dropped in the fall, after berry crops were past their prime. Some bears, especially females, became quite lethargic at this time.

In general, adult males were more mobile than females. Bears didn't use established trails, but used the same travel routes to move from one area of their home range to another. They sometimes ventured outside their traditional home range when food was scarce. In early spring, we found bears at lower elevations feeding on newly emerged grasses and forbs. As the snow melted at higher elevations, revealing new plant growth, the bears moved up. In late summer/early fall, they moved down again to eat ripe berries and remained at lower elevations until they denned.

Our Lowell study gave us the opportunity to study bear homing movements. At Lowell, 72 bears were moved to new locations in northern Idaho. Seven of these bears returned home to the Lowell area, suggesting that some bears have strong homing instincts and can move long distances over rough, mountainous terrain.

When we looked at home range use, we found that it varied according to age and sex. Newborn cubs stayed with their mothers for the first year. The following June, after the family broke up, most yearlings stayed within their mother's home range and denned there the following fall. During that time, male yearlings used larger areas than female yearlings. Adult males also occupied larger home ranges than adult females. By using home ranges that encompassed four to six female home ranges, males increased their odds of breeding each year. Subadult males used home ranges that were smaller than those of adult males but larger than those of adult females. Subadult females used home ranges about the same size as those of adult females. Male and female bears at Priest Lake, as well as females at Council, used the same home ranges from year to year. However, males at Council occasionally shifted their home ranges to take advantage of more plentiful foods, particularly berries.

Home ranges often overlapped, and bears didn't try to keep

other bears out of their home range. We believe this occurred because food supplies were patchy and unpredictable, so bears had to have large home ranges to get enough food. It was more advantageous for the bears to allow other bears in their home ranges than to spend energy defending those ranges. Female bears tended to minimize contact with nearby females by concentrating their activities in a small portion of their home range.

During our trapping studies, we saw that bears reacted to our presence in several ways. When we approached bears trapped in foot snares to inject them with immobilizing drugs, some moaned, growled, or tried to charge us. Others redirected their aggression from us to a nearby object. Some tried to run or climb a tree, while several covered their head with their paws or tried to hide behind a shrub or tree. In some cases, the bears sat with their head facing away from us or didn't move. Some bears that initially acted submissive or tried to flee became aggressive when we approached them to inject the immobilizing drugs. Adult males were more aggressive than subadult males, but we didn't see any behavioral difference between adult males and adult females.

HABITAT USE

Timber was the most frequently used habitat on the Council study area. It was especially important as a bedding site. Bears used open timber areas, selection cuts, and meadows in the spring, when they fed on grasses and forbs. In the summer and fall, shrubfields were important sources of berries. In general, bears in the Council area, especially females, stayed within 100 yards of water. They didn't use riparian areas as much as we expected, possibly because these areas were scarce and weather conditions were wet during the study.

Council bears rarely selected rock/talus, sagebrush/grass, and clear-cut habitats. These habitats supplied some bear foods, but other habitats apparently had more food and greater security cover. The bears also avoided roads. Although male bears used roads more often than females, overall both sexes preferred to stay more than 50 yards from roads, except when feeding. Females with cubs avoided both roads and open timber, which didn't provide sufficient cover. Most bears, except when they were feeding, stayed within 25 yards of cover.

Council bears preferred certain elevations, aspects, and topographic classes. The bears moved up and down in elevation depending on when their favorite foods were ripe. They started low in the spring, moved to higher elevations in the summer, and returned to lower elevations in late summer/early fall. The bears used steep, north-facing slopes for bedding sites and as travel corridors. They avoided ridgetops and upper slopes, preferring lower slopes, which were wetter than other areas and provided more food and cover.

Priest Lake bears liked diverse habitats. Wildfire and selection-cut logging created these habitats and increased the productivity and abundance of bear foods, especially berries. We found that selection cuts were the most important habitats used by bears because they provided not only plentiful food but also cover. Most selection cuts on the study area were 20 to 40 years old, and contained a dense shrub understory. Timber was the second-most used habitat at Priest Lake. Because bears made greater use of timber habitats in spring and fall, seasons in which they spent most of the day bedded, we suspect that timber functioned primarily as cover for bears. Female bears tended to use timber habitats more than males, perhaps because of the security cover the trees provided. Black bears used riparian areas, wet meadows, and avalanche chutes less than we expected. Riparian areas contained plants similar to those of adjacent forest habitats, so didn't offer bears much more in the way of food. Wet meadows and avalanche chutes were scarce and located at high elevations in inaccessible areas.

Bears avoided clear-cut areas and roads. Most clear-cuts were less than 20 years old and didn't provide adequate cover or the types of plant foods that bears preferred. Roads weren't popular, especially with females. Males, who were more mobile, used roads when they were available.

Based on our studies at Council and Priest Lake, we recommend that forest managers use logging techniques that minimize soil disturbance (which affects food production) and maintain bear security cover, bedding areas, and travel routes.

FOOD HABITS

Black bears are omnivorous but feed primarily on plants. Because they can't digest plant fibers, they don't extract many nutrients from plants. To survive, therefore, they must eat large amounts of food each day. The distribution and abundance of food affects the bears' growth rates, productivity, movements, and survival.

In the spring and early summer, bears ate less nutritious foods, such as grasses, forbs, and horsetails. Our data showed that some bears were able to maintain their body weight during this time; however, most bears lost weight. The bears' diet changed in late summer and fall as berries ripened. These fruits provided energy-rich sugars that allowed bears to put on body fat and were essential to bear reproduction and survival. Council bears ate an average of eight berry species, while Priest Lake bears had only three species available. Because of this, in years when one or more berry crops failed, the Priest Lake bears were more likely to starve or not produce young.

Insects were the most important animal food eaten by black bears. Between early and late summer, bears ate many social insects, especially ants. Ants also played a vital role in survival during drought years and when early season berry crops failed. Other animal foods such as birds, deer, and elk made up less than 2 percent of the bears' diet.

POPULATION CHARACTERISTICS

Our population characteristics studies indicate that hunting pressure can affect population age structures. In general, adult males were more vulnerable to hunters than younger bears. When hunting pressure was increased, the number of young males in a population rose and the number of adult males fell. The 1978 Coeur d'Alene population and the Lowell population were lightly hunted: In these populations, the median age was six years or older, and more than 60 percent of the males were adults. In the heavily hunted Council population and the 1983 Coeur d'Alene population, the median age was three years, and less than 50 percent of the males were adults. The movement of young bears through the study areas was partially responsible for the increase in young males we saw. These young bears had left their natal ranges in search of permanent home ranges.

Hunting pressure can also affect the sex composition of bear populations. Because adult males were more vulnerable to hunters than adult females, we expected male:female ratios in lightly hunted populations to favor adult males. We anticipated the opposite result in heavily hunted populations. In the lightly hunted Coeur d'Alene population (1978), we captured significantly more males than females. But at Lowell, another lightly hunted area, the male:female ratio was 50:50. We believe the trapping method we used at Lowell increased the number of females caught. We trapped more females than males in the heavily hunted Council population and caught fewer adult males in the heavily hunted Coeur d'Alene population (1983).

Our study of reproductive characteristics showed that Idaho black bears are long-lived, mature late, and have low reproductive rates. The breeding season for Idaho bears extends from mid-May until early August. After a gestation period of 220 days, cubs are born in late January or early February. The average age when females first bred was five years; no bears younger than four years old had cubs. This age of first reproduction was lower than those seen in Montana, Alaska, and Ontario but higher than those reported for the eastern United States. The average litter size was 1.7 cubs. Most of the females we monitored bred every other year, but some didn't breed for at least two years in a row. We found that diet and nutrition affected bear reproduction. In 1979, the fall huckleberry crop at Priest Lake was poor; the following year, none of the females we caught had cubs.

It was hard to get accurate estimates of bear population sizes and densities because black bears are secretive and live in forested areas. However, from our trapping studies, we estimated that 77 bears (1.5 bears/square mile) used the Council study area, 94 bears (0.8 bears/square mile) used the Priest Lake area, and 116 bears (1.2 bears/square mile) used the Lowell area.

Because trapping was expensive and time-consuming, it wasn't feasible for long-term monitoring of bear populations. In 1988–1990, we tested a bait station survey that was designed to measure population trends. We counted the number of bear visits to a series of scented bait stations placed in known bear habitats. We found that the survey results correlated well with the population data we collected from trapped bears.

Short-term changes in the size of bear populations were related to changes in the birth rate, which in turn were associated with the availability of nutritious foods, especially berries. Habitat quality and quantity, therefore, were important factors controlling population size because they influenced bear reproduction. The dispersal of young adults from their natal range didn't have an effect on population size unless the dispersing bears were forced to enter marginal or less secure habitats and died. Once bears became adults, very few died of natural causes. Most died during hunting seasons.

HARVEST CHARACTERISTICS

Data collected from the telephone harvest survey showed a 66 percent increase in bear harvest from 1983 to 1990. Most of this increase occurred in the spring bear hunting season, and we attributed it to greater hunting pressure. However, during this time, actual bear tag sales decreased. In spite of these conflicting trends, we believe that interest in bear hunting was increasing.

Hunters use four methods to take bears: still hunting, hound hunting, bait hunting, and incidental hunting. From 1983–1990, still hunters took 32 percent of the bears killed, incidental hunters took 29 percent, hound hunters took 20 percent, and bait hunters took 19 percent. Bait hunters experienced the largest increase in harvest during that period, followed by hound hunters and still hunters. Bait hunters and hound hunters had the highest success rates.

Between 1983 and 1990, rifle hunters killed the largest number of bears, followed by archers, and hunters using muzzleloaders and pistols.

The intensity of hunting pressure affects bear harvest rates and ultimately the number of bears in a population. As hunting pressure increases, more adult males (the most vulnerable members of a bear population) are harvested. If heavy hunting pressure continues, older males are depleted from the population and more young males and females are harvested. At even higher harvest levels, more females are taken. The population may then decline unless there are nearby reservoir areas, which produce young bears who can help repopulate heavily hunted areas. Bears are less vulnerable to hunters in areas with dense cover and/or few roads to provide hunters with access.

Our harvest data from 1983 to 1990 showed that on a statewide basis, bear harvest was occurring within acceptable limits. Yet some data suggested that the situation was changing. Interest in bear hunting was rising, the harvest was increasing, it was taking longer to kill a bear, and hunters were taking more females, especially older ones. Although we believe we were starting to see some symptoms of bear overharvest, we didn't see a significant population decline between 1983 and 1990.

Studies have shown that bear populations, because of their low reproductive rates, are slow to respond to hunting regulation changes designed to reduce harvest. Therefore, we suggest that bears be managed based on the risk of overharvest. If we wait until the evidence of overharvest is undeniable, we will have to severely cut hunting opportunity to recover bear populations. We recommend that bear managers establish bait station surveys to monitor populations; protect reservoir areas and movement corridors for dispersing bears; minimize habitat loss; conduct more bear studies; and adjust hunting seasons so more adult females will survive.

DENNING

Bears generally began denning in mid-October. Pregnant females usually denned before males. Before entering their dens each fall, female bears became less active and spent several days near their dens before actually entering them. Males usually denned soon after arriving at the site.

During hibernation, the bears' heart and breathing rates fell significantly, but their body temperatures dropped only slightly. We believe that black bears are more efficient hibernators than smaller mammals because they don't wake periodically to urinate, defecate, or feed, as many "true" hibernators do.

When we visited denned bears during their hibernation, they responded in several ways. As we approached the den site, most bears raised their heads and looked at us. Only rarely did they become aggressive. Some bears would "chomp" their jaws, growl at us, or swat their forepaws at us or the jab stick. These aggressive responses occurred most often early or late in the denning period, when the bears were less sleepy. Some of the bears we disturbed during the early stages of the denning period abandoned their dens.

Most of the bears we studied began to shed their foot pads late in the denning period. This shedding may have hampered bear movements immediately after they emerged from their dens, but the bears' lethargy after emergence probably restricted their movements more than any other factor.

Bears emerged from their dens in mid- to late April. We observed that females with new cubs were the last to leave their dens. In addition, bears who denned at lower elevations generally emerged before those who denned at higher elevations. After the bears left their den site, they usually moved to an area below the snow line and constructed day beds by scraping vegetation into a pile at the base of a tree. They stayed close to these day beds for several days.

During our denning studies, we described 45 ground and 11 tree dens. The dens of males and females didn't differ in size. Although males are larger than females, females with young needed dens as big as male dens. However, males dug larger den entrances. Several bears converted old coyote or badger dens into usable dens. Many of the bears lined their dens with nest materials, such as brush, conifer boughs, grass, or woodchips.

Most of the dens we located were ground dens dug into a hillside or under the base of a tree, stump, shrub, or fallen tree. We found other dens in the base of hollow trees or logs, or in rock cavities. Grand fir was the most common tree used for denning, followed by subalpine fir and Douglas fir. We found tree dens at higher elevations than ground dens, probably because grand fir and subalpine fir grow at high elevations. Dens at lower elevations were often located in dense brush pockets that provided cover, whereas high-elevation dens were often in relatively open areas, where accumulated snow provided protection. We found many of the dens in timber stands on north or west aspects. Although most bears preferred secure areas for denning, several bears denned less than 50 yards from well-used roads and some females denned in relatively open clear-cuts at higher elevations.

The bears apparently maintained several dens within their home ranges for hibernation. Some bears reused their dens if they weren't disturbed and the dens were durable. Yearlings, especially females, often used dens built by their mothers. Our data showed that yearlings who didn't learn denning behavior from their mothers were still able to successfully den. This indicated that denning behavior was instinctive rather than learned.

After analyzing our data, we concluded that the availability of dens wasn't limiting bear population size in the Council area. Most bears at Council didn't require large areas of undisturbed habitat for denning, but did prefer concealed den sites. Although logging on the Council study area reduced the number of suitable trees for denning, most bears could easily dig ground dens in the area's soils. However, in areas where bears can't easily dig in the soil, they may depend more on old-growth timber stands for denning sites. The dens we examined weren't concentrated in any core areas, so it wouldn't be feasible to manage specific locations as core denning areas. Perhaps this could be done in areas where bears extensively use trees for denning.

THE FUTURE OF IDAHO'S BLACK BEARS

As wildlife managers, we juggle many diverse issues in our attempts to integrate the needs and desires of humans with the biological needs of bears. The Idaho Department of Fish and Game periodically develops black bear management plans that serve as guidelines for setting bear hunting seasons and establish the department's philosophy and management direction for bears. Since 1980, three management plans have been written for Idaho's black bears. Each plan was a step forward in the quest to develop the best management program for black bears.

Predicting the future of Idaho's black bear populations, however, is a difficult task. Part of the difficulty lies in the fact that the department is undergoing a transition in terms of identifying and responding to its constituency – the people of Idaho. We can no longer make decisions based purely on what's biologically best for bears. We now have to consider sociological factors, such as changing human population demographics. For example, more Idahoans are now living in urban rather than rural settings.

As a result of these sociological trends, our bear management decisions have become controversial. Hunters and nonhunters are squaring off on issues such as spring bear hunting seasons, or using bait or hounds to hunt bears. Unfortunately, the public's attention has become focused on these issues, which have only short-term implications and minimal biological impact on bears. Habitat fragmentation and loss, a more important factor in the long-term survival of bear populations, is forgotten in these discussions.

Black bear populations are resilient and respond favorably to our management efforts. Bear hunting, in general, causes short-term population fluctuations that can be effectively dealt with by adjusting hunting seasons. Although no reliable methods are available to estimate black bear numbers, we do monitor population trends using a scent station survey and collect harvest data through the mandatory check program. These data help us compensate for the lack of precision we must deal with in managing wild bear populations.

Habitat loss, on the other hand, has a more subtle, yet permanent impact on the long-term viability of bear populations. Ultimately, the accelerating pace of habitat fragmentation and loss will dictate how long we can maintain bear populations. The prognosis for Idaho bears remains positive, because a majority of our land base is publicly owned and much of Idaho consists of rugged, mountainous terrain. As long as we continue to consider the well-being of bears and manage their habitat conservatively, we will continue to count the bear as a wild resident of Idaho.

APPENDIX I

COMMON AND SCIENTIFIC NAMES OF PLANTS, ANIMALS, PARASITES, AND DISEASE ORGANISMS

Common Name	Scientific Name
TREES	
Ponderosa pine	*Pinus ponderosa*
Douglas fir	*Pseudotsuga menziesii*
Hawthorn (black)	*Crataegus douglasii*
Red hawthorn	*Crataegus columbiana*
Chokecherry	*Prunus virginiana*
Elderberry	*Sambucus cerulea*
Grand fir	*Abies grandis*
Subalpine fir	*Abies lasiocarpa*
Engelmann spruce	*Picea engelmannii*
Bittercherry	*Prunus emarginata*
Dogwood	*Cornus stolonifera*
Mountain ash	*Sorbus scopulina*
Whitebark pine	*Pinus albicaulis*
Lodgepole pine	*Pinus contorta*
Western larch	*Larix occidentalis*
Western red cedar	*Thuja plicata*
White pine	*Pinus monticola*
Scouler willow	*Salix scoulerana*

Mountain maple	*Acer glabrum*
Serviceberry	*Amelanchier alnifolia*
Western hemlock	*Tsuga heterophylla*
Quaking aspen	*Populus tremuloides*

SHRUBS

Big sagebrush	*Artemisia tridentata*
Huckleberry	*Vaccinium globulare*
Buffaloberry	*Shepherdia canadensis*
Ceanothus (Shiny leaf)	*Ceanothus velutinus*
Red stem ceanothus	*Ceanothus sanguineus*
Twinberry	*Lonicera utahensis*
Syringa	*Philadelphus lewisii*
Ocean spray	*Holodiscus discolor*
Grouseberry	*Vaccinium scoparium*
Snowberry	*Symphoricarpos oreophilus*
Fool's huckleberry	*Menziesia ferruginea*
Ninebark	*Physocarpus malvaceus*
Mountain lover	*Pachistima myrsinites*

FORBS/GRASSES

Horsetail	*Equisetum* sp.
Clover	*Trifolium* sp.
Licoriceroot	*Ligusticum verticillatum*
Strawberry	*Fragaria* sp.
American false hellebore	*Veratrum viride*
Beargrass	*Xerophyllum tenax*
Pine grass	*Calamagrostis rubescens*

ANIMALS

Black bear	*Ursus americanus*
Elk	*Cervus elaphus*
Panda bear	*Ailuropoda melanoleuca*
Sun bear	*Helarctos malayanus*
Sloth bear	*Melursus ursinus*
Spectacled bear	*Tremarctos ornatus*
Etruscan bear	*Ursus etruscus*
Polar bear	*Ursus maritimus*
Cave bear	*Ursus spelaeus*
Brown bear (grizzly)	*Ursus arctos*

Mule deer	*Odocoileus hemionus*
Salmon	*Oncorhynchus tshawytscha*
Steelhead	*Salmo gairdneri*
Woodchuck	*Marmota monax*
Fox	*Vulpes* or *Urocyon* sp.
Coyote	*Canis latrans*
Badger	*Taxidea taxus*
Asiatic black bear	*Ursus thibetanus*
Hamster	*Cricetus cricetus*
Domestic pigeon	*Columba livia*

PARASITES AND DISEASE ORGANISMS

Tularemia	*Francisella tularensis*
Brucellosis	*Brucella abortus*
Leptospirosis	*Leptospira* spp.
Rocky Mountain wood tick	*Dermacentor andersoni*
American dog tick	*Dermacentor variabilis*
Lice	*Trichodectes pinguis enarctidos*
Fleas	*Chaetopsylla setosa*
Mites	*Ursicoptes americanus*
Toxoplasma	*Toxoplasma gondii*
Trichinella	*Trichinella spiralis*

APPENDIX 2

GLOSSARY OF TERMS

Acute lesion: A short-term skin inflammation, possibly accompanied by hair loss

Age of first reproduction: The age at which a male or female bear first produces young

Antibody: A body protein formed by exposure to a disease

Broadcast-burning: The practice of burning timber slash where it lies rather than bulldozing it into piles and then burning it

Bunodont: Dentition that is adapted to a diet of plant and animal foods

Caecum: The blind pouch in which the large intestine begins

Capture-mark-recapture study: A study in which bears are trapped, marked with ear tags, released and later recaptured

Chronic lesion: A long-term or recurring skin inflammation, possibly accompanied by hair loss

Circa-annual: Yearly

Clear-cut: A logging technique in which all of the trees in an area are cut

Cover type: A description of the plants or other natural features currently found on a site (for example, a meadow cover type contains predominantly grasses and forbs)

Cubby set: A log, brush or other enclosure used to funnel bears into a small area where they can be captured

Curvilinear: Represented by a curved line

Data Analysis Units (DAUs): Portions of bear management areas that were broken out so bear harvest data could be better analyzed. Each DAU has similar habitats, road access and proximity to urban population centers (for example, a DAU may have dense, semicoastal forest habitats, contain many roads and be close to cities).

Day bed: A temporary resting place used by bears when they aren't feeding

Dispersal: A behavior in which young bears leave their birth area to find a new home range

Disturbed vegetation types: Vegetation types that have been altered by human activities (for example, a clear-cut is a disturbed vegetation type)

Forb: A broad-leaved flowering plant

Habitat type: A description of the potential or climax vegetation on a site (for example, a Douglas fir/ninebark habitat type would be dominated by Douglas fir trees and ninebark shrubs when the site contained mature vegetation)

Harvest age structure: The percentage of young and old bears harvested (for example, if 70 out of 100 harvested bears were less than 4 years old, the harvest age structure would be 70 percent young bears and 30 percent adult bears)

Hunter effort: The average number of days a hunter spends pursuing his/her quarry (for example, hunters in one area may take an average of 10 days to kill a bear, while in another area, hunters may take an average of 15 days)

Intraspecific tolerance: How well members of the same species tolerate each other's presence

Jab stick: A six-foot-long stick with a syringe mounted on its end used to inject drugs into captured bears

Jackpot-burning: The practice of bulldozing timber slash into piles and then burning it

Leave patch or strip: A patch or strip of trees left standing in a logged area

Let-burn fire management policy: A policy in which wildfires are allowed to burn themselves out

Litter frequency: How often bears produce litters of cubs (for example, if one fifth of the females in a population had cubs each year, the average litter frequency would be 20 percent)

Mast: Berry or nut crops

Mesic vegetation: Plants that require a moderate amount of water

Natal range: The area where a bear is born

Periglaciated: Influenced by glacial forces

Plantigrade: Walking on the soles of the feet

Population age structure: The distribution of young and old bears in a population (for example, if 70 out of 100 bears in a population were less than 4 years old, the population age structure would be 70 percent young bears and 30 percent adult bears)

Potential climax series: A description of the mature plant community that could be supported on a site (for example, the western hemlock potential climax series would be dominated by western hemlock)

Prescribed burn: A controlled fire used by land managers to improve wildlife habitat

Radio-telemetry equipment: A radio transmitter placed around a bear's neck so that researchers can track the bear's movements

Reconstructed population: A minimum population size estimate obtained by adding the number of adult and subadult bears captured in one year to the number of unmarked bears captured in subsequent years that were alive the first year (for example, if 40 bears were captured the first year and 20

unmarked bears aged one year or older were captured the second year, the reconstructed population would be 60 bears)

Respiratory quotient: Breathing rate

Riparian: Located beside or near a watercourse, spring or other wet area

Scarification: Breaking up or disturbing the soil after logging has been completed

Scree: Rock-covered ground

Selection cut: A logging technique in which only the most valuable trees are cut

Seral: Plant species or communities that temporarily occupy a site. They will be replaced by other plant species or communities over time (for example, a plant community dominated by aspen may be slowly replaced by one dominated by Douglas fir)

Sex ratio: The ratio of male to female bears (for example, if 75 bears in a population were male and 25 bears were female, the sex ratio of males:females would be 75:25 or 3:1)

Sidehill park: An open area on a hillside

Sportsman Package: An Idaho license package that includes tags, permits and stamps for many Idaho fish and game species

Subadult: An immature bear less than four years old

Telemetered rectal temperature: A rectal temperature reading transmitted to researchers via radio signals

Titer: A test that shows whether a bear has been exposed to a disease

Triangulation: A method used by researchers to determine a bear's location by getting compass bearings from two or more spots. The point where the bearings cross is the approximate location of the bear.

Windrow-burning: The practice of bulldozing timber slash into rows and then burning it

BIBLIOGRAPHY

Ahlgren, C. 1966. Small mammals and reforestation following prescribed burning. J. For. 64:614–18.

Aldous, S. E. 1937. A hibernating black bear with cubs. J. Mammal. 18(4):466–68.

Alt, G. L. 1980. Rate of growth and size of Pennsylvania black bears. Pa. Game News. 51(12):7–17.

———, and J. J. Beecham. 1984. Reintroduction of orphaned black bear cubs into the wild. Wildl. Soc. Bull. 12:169–74.

———, G. J. Matula, Jr., F. W. Alt, and J. S. Lindzey. 1980. Dynamics of home range and movements of adult black bears in northeastern Pennsylvania. Int. Conf. Bear Res. and Manage. 5:131–36.

Amstrup, S. C., and J. J. Beecham. 1976. Activity patterns of radio-collared black bears in Idaho. J. Wildl. Manage. 40:340–48.

Asdell, S. A. 1964. Patterns of mammalian reproduction. Cornell University Press, Ithaca, N.Y. 670pp.

Barber, K. R., and F. G. Lindzey. 1986. Breeding behavior of black bears. Int. Conf. Bear Res. and Manage. 6:129–36.

Barnes, V. G., Jr., and O. E. Bray 1966. Black bears use drainage culverts for winter dens. J. Mammal. 47:712–13.

———, and O. E. Bray. 1967. Population characteristics and activities of black bears in Yellowstone National Park. Final Rep. Colo. Coop. Wildl. Res. Unit, Colorado State University,Fort Collins, 196pp.

Beecham, J. 1976. Black bear ecology. Job Prog. Rep. Idaho Dept. Fish and Game, Boise. 34pp.

———. 1977. Black bear ecology. Job Prog. Rep. Idaho Dept. Fish and Game, Boise, 43pp.

———. 1980. Population characteristics, denning, and growth patterns of black bears in Idaho. Ph.D. thesis, University of Montana, Missoula. 101pp.

———. 1982. Black bear ecology. Idaho Dept. Fish and Game Job Prog. Rep., Fed. Aid Proj. W-160-R-6.

———. 1983. Population characteristics of black bears in west central Idaho. J. Wildl. Manage. 47(2):402–12.

Beeman, L. E. 1975. Population characteristics, movements, and activities of the black bear (*Ursus americanus*) in the Great Smoky Mountains National Park. Ph.D. thesis, University of Tennessee, Knoxville. 218pp.

———, and M. R. Pelton. 1976. Homing of black bears in the Great Smoky Mountains National Park. Int. Conf. Bear Res. and Manage. 3:87–95.

———, and M. R. Pelton. 1980. Seasonal foods and the feeding ecology of black bears in the Smoky Mountains. Int. Conf. Bear Res. and Manage. 4:141–48.

Bennet, L. J., P. F. English, and R. L. Watts. 1943. The food habits of the black bear in Pennsylvania. J. Mammal. 24:24–31.

Bunnell, F. L., and D. E. N. Tait. 1981. Population dynamics of bears – Implications. Pages 75–98. In C. W. Fowler and T. D. Smith, eds. Dynamics of large mammal populations. John Wiley and Sons, Ltd., New York, N.Y. 477pp.

Caughley, G. 1974. Interpretation of age ratios. J. Wildl. Manage. 38:557–62.

———. 1977. Analysis of vertebrate populations. John Wiley and Sons, New York, N.Y. 23pp.

Cherry, J. S., and M. R. Pelton. 1976. Relationships between body measurements and weight of the black bear. J. Tenn. Acad. Sci. 51(1):32–34.

Collins, J. M. 1973. Some aspects of reproduction and age structures in the black bear in North Carolina. Proc. Ann. Conf. S.E. Assoc. Game and Fish Comm. 27:163–70.

Cowan, I. McT. 1938. Geographic distribution of color phases of the red fox and black bear in the Pacific Northwest. J. Marcum. 19(2):202–206.

———. 1972. The status and conservation of bears of the world – 1970. Int. Conf. Bear Res. and Manage. 2:343–67.

Craighead, F. C., Jr. 1971. Biotelemetry research with grizzly bears and elk in Yellowstone National Park, Wyoming, 1965. Pages 49–62. In Natl. Geogr. Soc. Res. Rep. 1965 Proj.

Craighead, J. J., and F. C. Craighead, Jr. 1972a. Grizzly bear prehibernation and denning activities as determined by radiotracking. Wildl. Monogr. No. 32. 35pp.

———, and F. C. Craighead, Jr. 1972b. Data on grizzly bear denning activities and behavior obtained by using wildlife telemetry. Int. Conf. Bear Res. and Manage. 2:84–106.

——, F. C. Craighead, Jr., J. R. Varney, and C. E. Cote. 1971. Satellite monitoring of black bear. BioScience. 21(24):1206–11.

——, J. R. Varney, F. C. Craighead, Jr., and J. S. Sumner. 1976. Telemetry experiments with a hibernating black bear. Int. Conf. Bear Res. and Manage. 3:357–71.

Daubenmire, R., and J. Daubenmire. 1968. Forest vegetation of eastern Washington and northern Idaho. Wash. Agric. Exp. Stn. Tech. Bull. 60. 104pp.

Egbert, A. L., and A. W. Stokes. 1976. The social behaviour of brown bears on an Alaskan salmon stream. In M. R. Pelton, J. W. Lentfer, and G. E. Folk, eds. Bears – Their biology and management. IUCN New Ser. 40:41–56.

Erickson, A. W. 1964. A mixed-age litter of brown bear cubs. J. Mammal. 45(2):312–13.

——. 1965. The brown-grizzly bear in Alaska. Its ecology and management. Fed. Aid Wildl. Restor. Proj. W-6-R-5, Work Plan F. Vol 5. Alaska Dept. Fish and Game, Juneau. 42pp.

——, and J. E. Nellor. 1964. Breeding biology of the black bear. Michigan State Univ. Agric. Exp. Stn. Res. Bull. 4:5–45.

——, and G. A. Petrides. 1964. Population structure, movements, and mortality of tagged black bears in Michigan. Mich. State Univ. Res. Bull. 4:46–67.

Eveland, J. F. 1973. Population dynamics, movements, morphology, and habitat characteristics of black bears in Pennsylvania. Pennsylvania State University, University Park. 157pp.

Fain, A. and D. E. Johnston. 1970. Un nouvel acarien de la famille Audycoptidae chez l'ours noir Ursus Americanus (sarcoptiformes). Acta Zool. Pathol. Antverpiensia. No. 50:179–81.

Farris, E. J., ed. 1950. The care and breeding of laboratory animals. John Wiley and Sons, Inc., New York, N.Y. 515pp.

Fendley, T. T., and I. L. Brisbin, Jr. 1977. Growth curve analyses: A potential measure of the effects of environmental stress upon wildlife populations. XIII Congr. Game Biol. 337–50.

Folk, G. E., Jr., A. Larson, and M. A. Folk. 1976. Physiology of hibernating bears. Int. Conf. Bear Res. and Manage. 3:373–80.

Franklin, J. F., and C. T. Dyrness. 1973. Natural vegetation of Oregon and Washington. U.S. For. Serv. Gen. Tech. PNW-8. 417pp.

Free, S. L., and E. McCafrey. 1972. Reproductive synchrony in the female black bear. Pages 199–206. In S. Herrero, ed. Bears – Their biology and management. IUCN Publ. New. Ser. 23.

Fuller, T. K., and L. B. Keith. 1980. Summer ranges, cover type use, and denning of black bears near Fort McMurray, Alberta, Canada. Field-Nat. 94:80–83.

Garshelis, D. L., and M. R. Pelton. 1980. Activity of black bears in the Great Smoky Mountains National Park. J. Mamm. 61(1):8–19.

Gershenson, S. 1945. Evolutionary studies on the distribution and dynamics of melanism in the hamster (*Cricetus cricetus* L.). I. Distribution of black hamsters in the Ukrainian and Bashkirian Soviet Socialist Republics (U.S.S.R.). Genetics. 30:207–32.

Goldsmith, A., M. E. Walraven, D. Graber, and M. White. 1981. Ecology of the black bear in Sequoia National Park. Natl. Park Serv. Final Rep. Contract No. CY-8000-4-0022. 64pp.

Graber, D. M. 1982. Ecology and management of black bears in Yosemite National Park. Final Rep. Natl. Park Serv., Yosemite Natl. Park. 206pp.

——, and M. White. 1983. Black bear food habits in Yosemite National Park. Int. Conf. Bear Res. and Manage. 5:1–10.

Greene, R. Idaho Dept. Lands, pers. commun.

Grenfell, W. E., and A. J. Brody. 1983. Seasonal foods of the black bears in Tahoe National Forest, California. California Fish and Game. 69:132–50.

Hagar, D. 1960. The interrelationships of logging, birds, and timber regeneration in the Douglas-fir region of northwestern California. Ecology. 41:116–25.

Hamer, D., and S. Herrero. 1987. Grizzly bear food and habitat in the front ranges of Banff National Park, Alberta. Int. Conf. Bear Res. and Manage. 7:199–213.

Hamilton, R. 1972. Summaries, by state: North Carolina. Pages 11–13. *In* R. L. Miller, ed. Proceedings of the 1972 black bear conference. N.Y. State Dept. of Environ. Conserv., Delmar, N.Y.

Hamilton, R. J., and R. L. Marchinton. 1980. Denning activity of black bears in the coastal plain of North Carolina. Pages 121–126. *In* C. J. Martinka and K. L. McArthur, eds. Bears – Their biology and management. 4:121–26.

Hammond, F. M. 1983. Food habits of black bears in the Greys River Drainage, Wyoming. M.S. thesis, University of Wyoming, Laramie. 50pp.

Hansson, A. 1947. The physiology of reproduction in mink (*Mustela vision* Schreb) with special references to delayed implantation. Acta Zool. 28:1–136.

Harlow, R. F. 1961. Characteristics and status of Florida black bear. Trans. North Am. Wildl. Conf. 26:481–95.

Harris, R. B. 1984. Harvest age structure as an indicator of grizzly bear population status. M.S. thesis, University of Montana, Missoula. 204pp.

Hatler, D. F. 1972. Food habits of black bears in interior Alaska. Can. Field-Nat. 86:17–31.

Hemmer, D. 1975. Serviceberry: Ecology, distribution, and relationships to big game. Montana Fish and Game Dept., Helena. Job Compl. Rep. Proj. W-120-R-5 and 6. 76pp.

Herrero, S. 1972. Aspects of evolution and adaptation in American black bears and brown and grizzly bears of North America. Int. Conf. Bear Res. and Manage. 2:221–31.

——. 1978. A comparison of some features of the evolution, ecology, and behavior of the black and grizzly/brown bears. Carnivore. 1:7–17.

Herrnstein, R. J. 1971. Quantitative hedonism. J. Psychiatric Res. 8:399–412.

Hock, R. J. 1951. Rectal temperatures of the black bear during its hibernation. Proc. Second Alas. Sci. Conf. 2:310–12.

——. 1960. Seasonal variation in physiological functions of arctic ground squirrels and black bears. Pages 155–171. In C. P. Lyman and A. R. Dawe, eds. Mammalian hibernation I. Harvard University, Cambridge, Mass..

Horn, H. S. 1968. The adaptive significance of colonial nesting in the Brewer's blackbird (Euphagus cyanocephalus). Ecology. 49(4):682–94.

Irwin, L. 1978. Relationships between intensive timber cultures, big game habitat, and elk habitat use patterns in northern Idaho. Ph.D. dissertation, University of Idaho, Moscow. 282pp.

Johnson, K. G. 1978. Den ecology of black bears in the Great Smoky National Park. M.S. thesis, University of Tennessee, Knoxville. 107pp.

——. 1982. Bait station surveys to determine relative density, distribution, and activities of black bears in the Southern Appalachian Region. Univ. of Tenn. Annu. Prog. Rep.

——, and M. R. Pelton. 1980. Environmental relationships and the denning period of black bears in Tennessee. J. Mammal. 61:653–60.

Jonkel, C. J. 1967. The ecology, population dynamics, and management of the black bear in the spruce-fir forests of northwestern Montana. Ph.D dissertation, University of British Columbia, Vancouver.

——, and I. M. Cowan. 1971. The black bear in the spruce-fir forest. Wildl. Monogr. No. 27. 55pp.

——, C. Lee, T. Thier, N. McMurray, and R. Mace. 1979. Sullivan Creek grizzlies. Border Grizzly Proj., Univ. Mont. Spec. Rep. No. 23. 10pp.

——, and F. L. Miller. 1970. Recent records of black bears on the barren grounds of Canada. J. Mammal. 51(4):826–28.

Kellyhouse, D. 1980. Habitat utilization by black bears in northern California. Int. Conf. Bear Res. and Manage. 4:221–27.

Kemp, G. A. 1972. Black bear population dynamics at Cold Lake, Alberta. 1968–1970. Pages 26–31. In S. Herrero, ed. Bears – Their biology and management. Int. Union Conserv. Nat. New Ser. 23. Morges, Switzerland.

——. 1976. The dynamics and regulation of black bear, Ursus americanus, populations in northern Alberta. Pages 191–197. In M. R. Pelton, J. W. Lentfer, and G. E. Folk, eds. Bears – Their biology and management. Int. Union Conserv. Nat. New Ser. 40. Morges, Switzerland.

——. 1979. Proceedings of the workshop on the management biology of North American black bear. *In* D. Burk, ed. The black bear in modern North America. The Amwell Press, N.J. 300pp.

Kingsley, M. C. S. 1979. Fitting the von Bertalanffy growth equation to polar bear age-weight data. Can. J. Zool. 57:1020–25.

——, J. A. Nagy, and R. H. Russell. 1983. Patterns of weight gain and loss for grizzly bears in northern Canada. Int. Conf. Bear Res. and Manage. 5:174–78.

Kohn, B. 1982. Status and management of black bears in Wisconsin. Tech. Bull. No. 129, Wisconsin Dept. Nat. Res. 31pp.

Kolenosky, G. B. 1986. The effects of hunting on an Ontario black bear population. Int. Conf. Bear Res. and Manage. 6:45–55.

Kordek, W. S., and J. S. Lindzey. 1980. Preliminary analysis of female reproductive tracts from Pennsylvania black bears. Int. Conf. Bear Res. and Manage. 4:159–62.

Landers, J. L., R. J. Hamilton, A. S. Johnson, and R. L. Marchinton. 1979. Foods and habitat of black bears in southeastern North Carolina. J. Wildl. Manage. 43:143–53.

Lawrence, W. 1979. Proceedings of the workshop on the management biology of North American black bear. *In* D. Burk, ed. The black bear in modern North America. The Amwell Press, N.J. 300pp.

LeCount, A. L. 1980. Some aspects of black bear ecology in the Arizona chaparral. Int. Conf. Bear Manage. and Res. 4:175–79.

——. 1982a. An analysis of the black bear harvest in Arizona (1968–1978). Arizona Game and Fish Dept. Spec. Rep. No. 12. Phoenix. 42pp.

——. 1982b. Characteristics of a central Arizona black bear population. J. Wildl. Manage. 46:861–68.

——, T. Waddell, and T. Beck, pers. commun.

Lentfer, J. W., R. J. Hensel, L. H. Miller, L. P. Glenn, and V. D. Berns. 1972. Remarks on denning habits of Alaska brown bears. Int. Conf. Bear Res. and Manage. 2:125–32.

Liche, H., and K. Wodzicki. 1939. Vaginal smears and the oestrous cycle of the cat and lioness. Nature. 144:245–46.

Lindzey, F. G. 1976. Black bear population ecology. Ph.D. thesis, Oregon State University, Corvallis. 105pp.

——, and E. C. Meslow. 1976a. Winter dormancy in black bears in southwestern Washington. J. Wildl. Manage. 40:408–15.

——, and E. C. Meslow. 1976b. Characteristics of black bear dens on Long Island, Washington. Northwest Sci. 50:236–42.

——, and E. C. Meslow. 1977. Home range and habitat use by black bears in southwestern Washington. J. Wildl. Manage. 41:413–25.

Ludlow, J. C. 1976. Observations on the breeding of captive black bears (*Ursus americanus*). Int. Conf. Bear Res. and Manage. 3:65–69.

Mace, R. D., and C. J. Jonkel. 1986. Local food habits of the grizzly bear in Montana. Int. Conf. Bear Res. and Manage. 6:105–10.

Maehr, D. S., and J. R. Brady. 1984. Food habits of Florida black bears. J. Wildl. Manage. 48(1):230–35.

———, and J. T. Defazio, Jr. 1985. Foods of black bears in Florida. Fla. Field-Nat. 13:8–12.

Manville, A. M. 1983. Human impact on the black bear population in Michigan's Lower Peninsula. Int. Conf. Bear Res. and Manage. 5:20–33.

Marcum, C. L., and D. O. Loftsgaarden. 1980. A nonmapping technique for studying habitat preferences. J. Wildl. Manage. 44:963–68.

Martin, P. 1979. Productivity and taxonomy of the *Vaccinium globulare, V. membranaceum* complex in western Montana. M.S. thesis, University of Montana, Missoula. 136pp.

———. 1983. Factors influencing globe huckleberry fruit production in northwestern Montana. Int. Conf. Bear Res. and Manage. 5:159–65.

Martinka, C. J. 1972. Habitat relationships of grizzly bears in Glacier National Park. Prog. Rep. 1972, Natl. Park Serv., Glacier Natl. Park, Mont. 19pp.

Matson, J. R. 1946. Notes on dormancy in the black bear. J. Mammal. 27(3):203–12.

———. 1954. Observations on the dormant phase of a female black bear. J. Mammal. 35(1):28–35.

McArthur, K. L. 1978. Homing behavior of transplanted black bears, Glacier National Park. Natl. Park Serv. Prog. Rep., Glacier Natl. Park, West Glacier, Mont. 24pp.

McIlroy, C. W. 1972. Effects of hunting on black bears in Prince William Sound. J. Wildl. Manage. 36:828–37.

Mealey, S. P. 1977. Method for determining grizzly bear habitat quality and estimating consequences of impacts on grizzly bear habitat quality. Final Draft. U.S. For. Serv. Reg. One. 36pp.

Miller, S. D. 1987. Susitna hydroelectric project. Final Rep. Big game studies: Black bear and brown bear. Alaska Dept. Fish and Game. 276pp.

———. 1990. Population management of bears in North America. Int. Conf. Bear Res. and Manage. 8:357–73.

———, and W. B. Ballard. 1982. Homing of transplanted Alaskan brown bears. J. Wildl. Manage. 46(4):869–76.

———, and S. M. Miller. 1988. Interpretation of bear harvest data. Alaska Dept. Fish and Game, Fed. Aid in Wildl. Restor. Proj. W-22-6, Job 4. 18 R. 65pp.

Minore, D. 1975. Observations on rhizomes and roots on *Vaccinium membranaceum*. U.S. For. Serv. Res. Note PNW-261. 5pp.

Mollohan, C. 1982. Black bear habitat research. Final Rep. No. 0182: 1 September 1982. Arizona Fish and Game, Phoenix.

Morrison, P. 1960. Some interrelations between weight and hibernation function. Bull. Mus. Comp. Zool. Harv. 124:75–91.

Mueggler, W. 1965. Ecology of seral shrub communities in the cedar-hemlock zone of northern Idaho. Ecol. Monogr. 35:165–85.

Mundy, K. R. D., and D. R. Flook, 1973. Background for managing grizzly bears in the national parks of Canada. Can. Wildl. Serv. Rep. Ser. 22. 35pp.

Mysterud, I. 1987. Bedding behavior in the European brown bear. In E. C. Meslow, ed. Bears – their biology and management.

Nelson, L. J. 1991. 1990 black bear opinion survey. Unpub. report, Idaho Dept. Fish and Game. 8pp.

Nelson, R. A., N. W. Wahner, J. D. Jones, R. D. Ellefson, and P. E. Zollman. 1973. Metabolism of bears before, during, and after winter sleep. Am. J. Physiol. 224(2):491–96.

Novick, H. J., and G. R. Stewart. 1982. Home range and habitat preferences of black bears in the San Bernardino Mountains of southern California. California Fish and Game. 67:21–35.

Payne, N. F. 1975. Unusual movements of Newfoundland black bears. J. Wildl. Manage. 39(4):812–13.

Pearson, A. M. 1975. The northern interior grizzly bear Ursus arctos. Can. Wildl. Serv. Rep. Ser. No. 34, Ottawa. 86pp.

——. 1976. Population characteristics of the arctic mountain grizzly bear. Pages 247–260. In M. R. Pelton, J. W. Lentfer, and G. E. Folk, eds. Bears – Their biology and management. Int. Union Conserv. Nat. New Ser. 40. Morges, Switzerland.

Pelton, M. R., L. E. Beeman, and D. C. Eagar. 1980. Den selection by black bears in the Great Smoky Mountains National Park. Int. Conf. Bear Res. and Manage. 4:149–51.

Pfister, R., B. Kovalchik, S. Arno, and R. Presby. 1977. Forest habitat types of Montana. U.S. For. Serv. Gen. Tech. Rep. INT-34. 174pp.

Piekielek, W., and T. S. Burton. 1975. A black bear population study in northern California. California Fish and Game. 61(1):4–25.

Poelker, R. J., and H. D. Hartwell. 1973. Black bear of Washington. State Game Dep. Biol. Bull. 14. 180pp.

Quinn, P. J., R. O. Ramsden, and D. H. Johnston. 1976. Toxoplasmosis: A serological survey in Ontario wildlife. J. Wildl. Dis. 12:504–10.

Raine, R. M., and J. L. Kansas. 1990. Black bear seasonal food habits and distribution by elevation in Banff National Park. Alberta Int. Conf. Bear Res. and Manage. 8:297–304.

Rausch, R. L. 1961. Notes on the black bear (Ursus americanus Passas) in Alaska, with particular reference to dentition and growth. Z.f. Saugetierkunde. 26:65–128.

Raybourne, J. W. 1976. A study of black bear populations in Virginia. Trans. Northeast. Sect., The Wildl. Soc., Fish and Wildl. Conf. 33:71–81.

Reynolds, D. G., and J. J. Beecham. 1980. Home range activities and reproduction of black bears in west central Idaho. Int. Conf. Bear Res. and Manage. 4:181–90.

Reynolds, H. V., J. A. Curatolo, and R. Quimby. 1976. Denning ecology of grizzly bears in northeastern Alaska. Int. Conf. Bear Res. and Manage. 3:403–409.

Ricklefs, R. E. 1968. Patterns of growth in birds. Ibis 110:419–51.

Robinette, W. L., C. H. Baer, R. E. Pillmore, and C. E. Knittle. 1973. Effects of nutritional change on captive mule deer. J. Wildl. Manage. 37:312–26.

Rogers, L. L. 1974. Shedding of foot pads by black bears during denning. J. Mammal. 55(3):672–74.

———. 1976. Effects of mast and berry crop failures on survival, growth, and reproductive success for black bears. Proc. North Am. Wildl. and Nat. Resour. Conf. 41:431–38.

———. 1977. Social relationships, movements, and population dynamics of black bears in northeastern Minnesota. Ph.D. thesis, University of Minnesota, Minneapolis. 194pp.

———. 1978. Effects of food supply, predation, cannibalism, parasites, and other health problems on black bear populations. Bunnell, Eastman, and Peek, eds. Symp. Sat. Reg. Wildl. Populations. Vancouver, B.C.

Rohlman, J. A. 1989. Black bear ecology near Priest Lake, Idaho. M.S. thesis, University of Idaho, Moscow. 76pp.

Russell, R. H., J. W. Nolan, N. G. Woody, G. H. Anderson, and A. M. Pearson. 1978. A study of the grizzly bear in Jasper National Park: A progress report 1976 and 1977. Prep. for Parks Canada. Prep. by Can. Wildl. Serv., Edmonton, Alberta. 95pp.

Sadleir, R. M. F. S. 1969. The role of nutrition in the reproduction of wild mammals. J. Reprod. Fer. 6:39–48.

Sauer, P. R. 1975. Relationships of growth characteristics to sex and age for black bears from the Adirondacks region of New York. New York Fish and Game J. 22(2):81–113.

Schwartz, C. C., A. W. Franzmann, and D. C. Johnson. 1983. Black bear predation on moose. Final Rep. Alaska Dept. Fish and Game, Juneau. 135pp.

Serveheen, C. 1981. Grizzly bear ecology and management in the Mission Mountains, Montana. Ph.D. dissertation, University of Montana, Missoula. 139pp.

———, and R. Klaver. 1983. Grizzly bear dens and denning activity in the Mission and Rattlesnake mountains, Montana. Int. Conf. Bear Res. and Manage. 5:203–209.

Shaffer, S. 1971. Some ecological relationships of grizzly bears and black bears of the Apgar Mountains in Glacier National Park, Montana. M.S. thesis, University of Montana, Missoula. 134pp.

Spencer, H. E., Jr. 1955. The black bear and its status in Maine. Maine Dept. Inland Fish and Game Bull. 4. 55pp.

Standley, P. C. 1921. Albinism in the black bear. Sci. 54 (1386):74.

Stickley, A. R., Jr. 1961. A black bear tagging study in Virginia. Proc. Ann. Conf. S.E. Game and Fish Comm. 15:43–54.

Stoneberg, R. P., and C. J. Jonkel. 1966. Age determination of black bears by cementum layers. J. Wildl. Manage. 30(2):411–14.

Stonorov, D. S., and A. W. Stokes. 1972. Social behavior of the Alaskan brown bear. Int. Conf. Bear Res. and Manage. 2:232–42.

Stringham, S. F. 1980. Possible impacts of hunting on the grizzly/brown bear, a threatened species. Pages 337–349. In C. J. Martinka and K. L. McArthur, eds. Bears – Their biology and management. U.S. Gov. Printing Off., Washington, D.C.

Svihla, A. H., and H. S. Bowman. 1954. Hibernation in the American black bear. Am. Mid. Nat. 52(1):248–52.

Tisch, E. L. 1961. Seasonal food habits of the black bear in the Whitefish Range of northwestern Montana. M.S. thesis, Montana State University, Missoula. 108pp.

Tizard, I. R., J. B. Billett, and R. O. Ramsden. 1976. The prevalence of antibodies against Toxoplasma gondii in some Ontario mammals. J. Wildl. Dis. 12:322–25.

Troyer, W., and J. B. Faro. 1975. Aerial survey of brown bear denning in the Katmai area of Alaska. Pres. at Northwest Sect. Wildl. Soc. Meet. 2–4 Apr. 1975, Anchorage, Alas. 10pp.

Unsworth, J. W., J. J. Beecham, and L. R. Irby. 1989. Female black bear habitat use in west central Idaho. J. Wildl. Manage. 53(3):668–73.

Vaughan, M. R., E. J. Jones, D. W. Carney, and N. Garner. 1983. Seasonal habitat use and home range of black bears in Shenandoah National Park: First Ann. Prog. Rep. 1 July 1983.

Vroom, G. W., S. Herrero, and R. T. Ogilvie. 1980. The ecology of winter den sites of grizzly bears in Banff National Park, Alberta. Int. Conf. Bear Res. and Manage. 4:321–30.

Watson, A., and R. Moss. 1971. Spacing as affected by territorial behavior, habitat, and nutrition in red grouse (Lagopus lagopus scoticus). Pages 92–111. In A. H. Esser, ed. The use of space by animals and men. Plenum Press, N.Y. 411pp.

Weiner, J. G., and S. E. Fuller. 1975. Comments on "Environmental evaluation based on relative growth rates of fishes." Prog. Fish Culturist. 37(2):99–100.

Weins, J. A. 1976. Population responses to patchy environments. Annu. Rev. Ecol. Syst. 7:81–120.

West, N. E., and R. W. Wein. 1971. A plant phenological index. BioScience. 21(3):116–17.

Willey, C. H. 1972. Vulnerability of bears to hunting. Pages 24–27. In R. L. Miller, ed. Proceedings of the 1972 black bear conference. N.Y. State Dept. Environ. Conserv. Delmar, N.Y.

———. 1978. The Vermont black bear. Vermont Fish and Game Dept., Montpelier. 73pp.

Wimsatt, W. A. 1963. Delayed implantation in the Ursidae, with particular reference to the black bear (Ursus americanus Pallus). Pages 49–76. In A. C. Enders, ed. Delayed implantation. The University of Chicago Press, Chicago, Ill.

Wood, A. J., I. McT. Cowan, and H. C. Nordan. 1962. Periodicity of growth in ungulates as shown by deer of the genus Odocoileus. Can. J. Zool. 40:593–603.

Worley, D. E., J. C. Fox, J. B. Winters, and K. R. Greer. 1974. Prevalence and distribution of Trichinella spiralis in carnivorous mammals in the U.S. Northern Rocky Mountain Region. In C. W. Kim, ed. Trichinellosis, Proc. Third International Conference on Trichinellosis. Intext Educational Publishers, New York, N.Y.

Young, B. F., and R. L. Ruff. 1982. Population dynamics and movements of black bears in east central Alberta. J. Wildl. Manage. 46(4):845–60.

Young, D. D. 1984. Black bear habitat use at Priest Lake, Idaho. M.S. thesis, University of Montana, Missoula. 66pp.

———, and J. J. Beecham. 1986. Black bear habitat use at Priest Lake, Idaho. Int. Conf. Bear Res. and Manage. 6:73–80.

Yunker, C. E., C. E. Binninger, J. E. Keirans, J. J. Beecham and M. Schlegel. Clinical mange in the black bear associated with Ursicoptes americanus (Acari: Audycoptidae). J. Wildl. Dis. 16(3):347–56.

Zager, P. E. 1980. The influence of logging and wildfire on grizzly bear habitat in northwestern Montana. Ph.D. dissertation, University of Montana, Missoula. 131pp.

———. 1981. Northern Selkirk Mountains grizzly bear habitat survey, 1981. U.S. For. Serv., Id. Panhandle Natl. For. Contract. 75pp.

———, and C. J. Jonkel. 1983. Managing grizzly bear habitat in the northern Rocky Mountains. J. For. 81(8):524–26.

———, C. Jonkel, and J. Habeck. 1983. Logging and wildfire influence on grizzly bear habitat in northwestern Montana. Int. Conf. Bear Res. and Manage. 5:124–32.

———, C. Jonkel, and R. Mace. 1980. Grizzly bear habitat terminology. Border Grizzly Proj., University of Mont., Missoula, Spec. Rep. No. 41. 15pp.

Zimmerman, W. J. 1977. Trichinosis in bears of western and north-central United States. Am. J. Epidemiol. 106:167–71.

INDEX

shy, 1, 13, 112, 129, 199, 207
study method, 58
territorial (*see* Home range
  overlap)
at trap site, 57–58, 69–71, 111,
  204, pl. 4
Berry crop. *See also* specific berry
  species
  food source, 89, 104, 105, 129,
    203, 206
  production, 80, 88, 109, 204,
    205
  effects on reproduction/
    survival, 1, 107, 129, 133, 206,
    207, 208
Big sage, 15
Biting, 174, 189
Bitter cherry, 15, 17, 80, 87,
  106
Breastfeeding. *See* Lactation
Breathing. *See* Physiology
Breeding. *See* Reproduction
Brown bear
  behavior, 41, 71, 191
  denning, 178, 180, 182–83, 185,
    186, 187
  distribution and evolution, 9,
    10, 199
  food resources and habitat, 88,
    89, 93
  home range overlap, 67
  homing movements, 63
  hunting pressure, 119
  mortality, 188
  physical characteristics, 36, 38,
    42
Buffaloberry, 15, 60, 80, 87, 106,
  107

Bureau of Land Management, 10
Capturing. *See* Trapping
Carnivore, 7, 8
Carrying capacity. *See* Habitat
Cave bear, 9
Ceanothus, 17
Characteristics, physical, 33–52,
  201–2
  description, 36
  length, 36, 38 (table 5–2), 40
  size, 33–34, 36, 42, 43, 44, 201
  study method, 34
  weight, 34, 36, 38 (table 5–2),
    39–40, 44, 201
  estimating, 33–34, 35 (table
    5–1)
  at sexual maturity, 43, 202
  variation, 36–42, 105–6,
    201–2, 206
Charging. *See* Behavior: aggressive
Chest marking. *See* Color phases
Chokecherry, 15, 60, 80, 87, 106,
  107
Classification, 7
Clear-cut. *See* Habitat
Clover, 17, 109
Coeur d'Alene study area, 5,
  17–18, 199–200
Coeur d'Alene study area bears
  age structure, 113 (table 9–1),
    117–20, 206–7
  bait station survey, 131–32
  characteristics, physical, 38
    (table 5–2), 47 (table 5–9),
    48 (table 5–10)
  color phases, 49, 202
  hunting pressure, 118, 120, 123,
    124–25, 206–7

types, 175 (table 11–1), 179–80,
210–11
ground, 179–80, 184 (table
11–4), 185–86, 193, 210–11,
pl. 9
other, 180, 186, 211
tree, 180, 184 (table 11–4),
185 (fig. 11–2), 186, 193,
210–11
use, 186–88, 211
Denning, 173–94, 209–11. *See also*
Hibernation
abandoning den, 177, 189, 190
(fig. 11–3), 210
behavior, 188, 211 (*see also*
Behavior: at den site)
chronology, 176–79, 209–10
evolution of, 178
habitats (*see* Habitat use
pattern)
management considerations,
193–94, 211
physiology, 38, 174, 189,
191–92, 210
sharing den, 188
study method, 23–24, 29,
32n.13, 173–75, 188, 201
Density. *See* Population
Dentition, 26, 36, 37 (fig. 5–1)
Depredation, 61, 107, 166
Description. *See* Characteristics,
physical
Devil's club berries, 60
Diet. *See* Food habits
Digestion, 104, 107, 206
Disease, 50–51, 202
Dispersal (subadult movements)

guidelines for aiding, 96, 171,
209
effects on home range size, 63,
66
and hunting pressure, 121–22,
151–52, 170, 209
effects on population size,
114–15, 116, 117, 134, 135, 208
reason for, 134–35, 207
Distribution, 1–2, 7, 8 (fig. 2–1),
9–10, 11 (fig. 2–3), 76, 199
Dogwood, 15
Dormancy. *See* Hibernation
Douglas fir, 15, 16, 17, 180, 211

Elderberry, 15
Elk River study area, 5, 19–20,
199–200
Elk River study area bears, 12, 113
(table 9–1), 131 (table 9–10)
Engelmann spruce, 15, 17
Environment. *See* Habitat
Environmental toxin. *See*
Mercury level; Pesticide level
Etruscan bear, 9
Evolutionary history, 7–9, 199

Family group, 64–65, 203
Feces. *See* Scat
Feeding. *See* Food habits
Fire, 87–89, 94, 200, 205
Food habits, 103–9, 206. *See also*
Behavior: feeding
animal/insect foods, 1, 87,
104, 105, 107–8, 206
late summer/fall, 86–87, 89,
104, 106–7, 206, pl. 7

age of first reproduction
Silviculture. *See* Logging
Size. *See* Characteristics, physical
Sleep pattern. *See* Bedding
Sloth bear, 7
Snowberry, 17
Social organization, 41–42, 63–69.
    See also Population age structure
Spectacled bear, 7, 8
Strawberry, 17
Study area. *See* Idaho Department
    of Fish and Game: bear
    research program
Study method. *See* Idaho
    Department of Fish and
    Game: bear research program
Subalpine fir, 15, 17, 180, 211
Sun bear, 7
Syringa, 17

Tag, ear, 25
Taxonomy, 7
Telemetry. *See* Radio-telemetry
Telephone harvest survey. *See*
    Harvest
Temperature. *See* Physiology
Territory. *See* Home range
Timber harvest. *See* Logging
Timber management. *See*
    Logging
Timber production. *See* Logging
Tooth. *See* Dentition
Tranquilizing, pl. 1
    at den site, 23–24, 29, 173–74,
    188, 189, 201

drugs, 25, 31nn.1–4
    at trap site, 25, 58, 69–70, 200,
    204
Trapping
    baits used, 24–25, 108
    releasing bears, 27, 57
    study method, 24–25, 111, 124,
    130, 200
    success, 14, 111–12, 113, 120, 124,
    127–28
Travel. *See* Movement
Tremarctinae, 8
Tremarctos, 7
Twinberry, 17, 80

U.S. Forest Service, 10, 16, 17, 95
Ursavus, 8
Ursid family (Ursidae), 7–8, 9,
    199
Ursinae, 8–9
Ursus, 1, 7, 8–9

Vegetation. *See* specific species
Vegetative zone, 17
Vocalizing. *See* Behavior:
    aggressive

Weight. *See* Characteristics,
    physical
Western larch, 16
Western red cedar, 16, 17
Whitebark pine, 15–16
White pine, 16, 17, 18
Wildfire. *See* Fire
Willow, 17